Enzyme Biotechnology

Edited by
Benjy Elliott

Larsen & Keller
www.larsen-keller.com

Enzyme Biotechnology
Edited by Benjy Elliott
ISBN: 978-1-63549-115-9 (Hardback)

© 2017 Larsen & Keller

 Larsen & Keller

Published by Larsen and Keller Education,
5 Penn Plaza,
19th Floor,
New York, NY 10001, USA

Cataloging-in-Publication Data

Enzyme biotechnology / edited by Benjy Elliott.
    p. cm.
Includes bibliographical references and index.
ISBN 978-1-63549-115-9
1. Enzymes. 2. Enzymes--Biotechnology. 3. Enzymology.
I. Elliott, Benjy.
QP601 .E59 2017
572.7--dc23

The publisher's policy is to use permanent paper from mills that operate a sustainable forestry policy. Furthermore, the publisher ensures that the text paper and cover boards used have met acceptable environmental accreditation standards.

Printed and bound in the United States of America.

For more information regarding Larsen and Keller Education and its products, please visit the publisher's website www.larsen-keller.com

Table of Contents

Permissions

Index

Preface

The book aims to shed light on some of the unexplored aspects of enzymes. It unfolds the innovative aspects of this field which will be crucial for the holistic understanding of the subject matter. Enzymes are very important for metabolic processes and human survival. They are macromolecular biological catalysts and are responsible for the acceleration of chemical reactions. The topics included in this book on enzymes are of utmost significance and bound to provide incredible insights to readers. Also included in this book is a detailed explanation of the various concepts and applications of enzyme biotechnology. It will serve as a valuable source of reference for those interested in this field.

To facilitate a deeper understanding of the contents of this book a short introduction of every chapter is written below:

Chapter 1- Enzymes are biological macromolecules that act as catalysts, converting substrates in metabolic processes to products. The chapter defines what an enzyme is, traces its etymology and provides an overview of its importance. It offers an insightful focus on enzymes, keeping in mind the complex subject matter.

Chapter 2- Enzymes are classified according to the way they work on a molecular level into categories like- transferases, isomerases etc. According to the site of their activity they can be divided into two categories- endoenzyme and exoenzyme. This chapter studies the categories comprehensively with suitable examples to help the reader understand the intricacies of each category.

Chapter 3- This chapter discusses the topics of enzyme inhibitors, enzyme activators and the phenomena of enzyme promiscuity. Enzyme inhibitors are molecules that adhere to enzymes and limit their activity while enzyme activators are molecules that attach to enzymes and augment their activity. These topics have been discussed in detail and a section of the chapter studies enzyme promiscuity the ability of an enzyme to catalyze a fortuitous side reaction in addition to its main reaction.

Chapter 4- When a chemical reaction involving enzymes undergoes an increase in its rate, the phenomenon is said to be an enzyme catalysis. This chapter studies the mechanism of action and the processes involved in enzyme catalysis by using enzyme kinetics. The reader gains a thorough understanding of the way enzymes act under varying conditions.

Chapter 5- This chapter focuses on the metabolic pathways of oxidative phosphorylation, citric acid cycle, glycolysis, pentose phosphate pathway, fatty acid synthesis etc. Understanding the metabolic pathways helps grasp how an enzyme functions, the series of reactions that take place by the action of an enzyme and the facilitators of enzyme reactions. The topics discussed in the chapter are of great importance to broaden the existing knowledge on enzymes.

Chapter 6- There are several causes behind enzyme depletion; this chapter studies the causes and the effects of enzyme depletion in detail. The chapter contains topics like chronic granulomatous disease, cortisone reductase deficiency, hypophosphatasia, myeloperoxidase deficiency, neutrophil-specific granule deficiency, phenylketonuria and pseudocholinesterase deficiency among others. The major components of enzyme deficiency are discussed in this chapter.

I would like to share the credit of this book with my editorial team who worked tirelessly on this book. I owe the completion of this book to the never-ending support of my family, who supported me throughout the project.

Editor

Introduction to Enzyme

Enzymes are biological macromolecules that act as catalysts, converting substrates in metabolic processes to products. The chapter defines what an enzyme is, traces its etymology and provides an overview of its importance. It offers an insightful focus on enzymes, keeping in mind the complex subject matter.

Enzymes are macromolecular biological catalysts. Enzymes accelerate, or catalyze, chemical reactions. The molecules at the beginning of the process upon which enzymes may act are called substrates and the enzyme converts these into different molecules, called products. Almost all metabolic processes in the cell need enzymes in order to occur at rates fast enough to sustain life. The set of enzymes made in a cell determines which metabolic pathways occur in that cell. The study of enzymes is called *enzymology*.

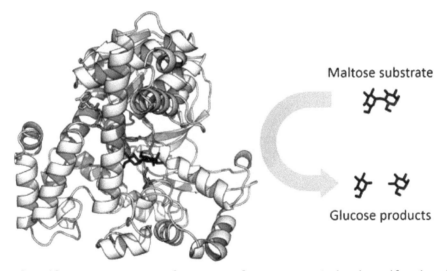

Maltose substrate

Glucose products

The enzyme glucosidase converts sugar maltose to two glucose sugars. Active site residues in red, maltose substrate in black, and NAD cofactor in yellow. (PDB 1OBB)

Enzymes are known to catalyze more than 5,000 biochemical reaction types. Most enzymes are proteins, although a few are catalytic RNA molecules. Enzymes' specificity comes from their unique three-dimensional structures.

Like all catalysts, enzymes increase the rate of a reaction by lowering its activation energy. Some enzymes can make their conversion of substrate to product occur many millions of times faster. An extreme example is orotidine 5'-phosphate decarboxylase, which allows a reaction that would otherwise take millions of years to occur in milliseconds. Chemically, enzymes are like any catalyst and are not consumed in chemical reactions, nor do they alter the equilibrium of a reaction. Enzymes differ from most other catalysts by being much more specific. Enzyme activity can be affected by other molecules inhibitors are molecules that decrease enzyme activity, and activators are molecules that increase activity. Many drugs and poisons are enzyme inhibitors. An enzyme's activity decreases markedly outside its optimal temperature and pH.

Some enzymes are used commercially, for example, in the synthesis of antibiotics. Some household products use enzymes to speed up chemical reactions enzymes in biological washing powders break down protein, starch or fat stains on clothes, and enzymes in meat tenderizer break down proteins into smaller molecules, making the meat easier to chew.

Etymology and History

By the late 17th and early 18th centuries, the digestion of meat by stomach secretions and the conversion of starch to sugars by plant extracts and saliva were known but the mechanisms by which these occurred had not been identified.

Eduard Buchner

French chemist Anselme Payen was the first to discover an enzyme, diastase, in 1833. A few decades later, when studying the fermentation of sugar to alcohol by yeast, Louis Pasteur concluded that this fermentation was caused by a vital force contained within the yeast cells called "ferments", which were thought to function only within living organisms. He wrote that "alcoholic fermentation is an act correlated with the life and organization of the yeast cells, not with the death or putrefaction of the cells."

In 1877, German physiologist Wilhelm Kühne (1837–1900) first used the term *enzyme*, which comes to describe this process. The word *en-zyme* was used later to refer to nonliving substances such as pepsin, and the word *fer-ment* was used to refer to chemical activity produced by living organisms.

Eduard Buchner submitted his first paper on the study of yeast extracts in 1897. In a series of experiments at the University of Berlin, he found that sugar was fermented by yeast extracts even when there were no living yeast cells in the mixture. He named the enzyme that brought about the fermentation of sucrose "zymase". In 1907, he received the Nobel Prize in Chemistry for "his discovery of cell-free fermentation". Following Buchner's example, enzymes are usually named according to the reaction they carry out the suffix *-ase* is combined with the name of the substrate (e.g., lactase is the enzyme that cleaves lactose) or to the type of reaction (e.g., DNA polymerase forms DNA polymers).

The biochemical identity of enzymes was still unknown in the early 1900s. Many scientists observed that enzymatic activity was associated with proteins, but others (such as Nobel laureate Richard Willstätter) argued that proteins were merely carriers for the true enzymes and that proteins *per se* were incapable of catalysis. In 1926, James B. Sumner showed that the enzyme urease was a pure protein and crystallized it; he did likewise for the enzyme catalase in 1937. The conclusion that pure proteins can be enzymes was definitively demonstrated by John Howard Northrop and Wendell Meredith Stanley, who worked on the digestive enzymes pepsin (1930), trypsin and chymotrypsin. These three scientists were awarded the 1946 Nobel Prize in Chemistry.

The discovery that enzymes could be crystallized eventually allowed their structures to be solved by x-ray crystallography. This was first done for lysozyme, an enzyme found in tears, saliva and egg whites that digests the coating of some bacteria; the structure was solved by a group led by David Chilton Phillips and published in 1965. This high-resolution structure of lysozyme marked the beginning of the field of structural biology and the effort to understand how enzymes work at an atomic level of detail.

Structure

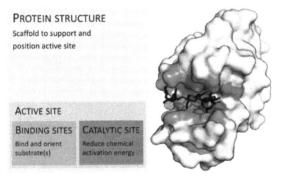

Organisation of enzyme structure and lysozyme example. Binding sites in blue, catalytic site in red and peptidoglycan substrate in black. (PDB 9LYZ)

Enzymes are generally globular proteins, acting alone or in larger complexes. Like all proteins, enzymes are linear chains of amino acids that fold to produce a three-dimensional structure. The sequence of the amino acids specifies the structure which in turn determines the catalytic activity of the enzyme. Although structure determines function, a novel enzyme's activity cannot yet be predicted from its structure alone. Enzyme structures unfold (denature) when heated or exposed to chemical denaturants and this disruption to the structure typically causes a loss of activity. Enzyme denaturation is normally linked to temperatures above a species' normal level; as a result, enzymes from bacteria living in volcanic environments such as hot springs are prized by industrial users for their ability to function at high temperatures, allowing enzyme-catalysed reactions to be operated at a very high rate.

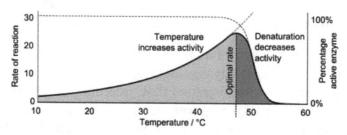

Enzyme activity initially increases with temperature (Q10 coefficient) until the enzyme's structure unfolds (denaturation), leading to an optimal rate of reaction at an intermediate temperature.

Enzymes are usually much larger than their substrates. Sizes range from just 62 amino acid residues, for the monomer of 4-oxalocrotonate tautomerase, to over 2,500 residues in the animal fatty acid synthase. Only a small portion of their structure (around 2–4 amino acids) is directly involved in catalysis the catalytic site. This catalytic site is located next to one or more binding sites where residues orient the substrates. The catalytic site and binding site together comprise the enzyme's active site. The remaining majority of the enzyme structure serves to maintain the precise orientation and dynamics of the active site.

In some enzymes, no amino acids are directly involved in catalysis; instead, the enzyme contains sites to bind and orient catalytic cofactors. Enzyme structures may also contain allosteric sites where the binding of a small molecule causes a conformational change that increases or decreases activity.

A small number of RNA-based biological catalysts called ribozymes exist, which again can act alone or in complex with proteins. The most common of these is the ribosome which is a complex of protein and catalytic RNA components.

Mechanism

Substrate Binding

Enzymes must bind their substrates before they can catalyse any chemical reaction. Enzymes are usually very specific as to what substrates they bind and then the chemical reaction catalysed. Specificity is achieved by binding pockets with complementary shape, charge and hydrophilic/hydrophobic characteristics to the substrates. Enzymes can therefore distinguish between very similar substrate molecules to be chemoselective, regioselective and stereospecific.

Some of the enzymes showing the highest specificity and accuracy are involved in the copying and expression of the genome. Some of these enzymes have "proof-reading" mechanisms. Here, an enzyme such as DNA polymerase catalyzes a reaction in a first step and then checks that the product is correct in a second step. This two-step process results in average error rates of less than 1 error in 100 million reactions in high-fidelity mammalian polymerases. Similar proofreading mechanisms are also found in RNA polymerase, aminoacyl tRNA synthetases and ribosomes.

Conversely, some enzymes display enzyme promiscuity, having broad specificity and acting on a range of different physiologically relevant substrates. Many enzymes possess small side activities which arose fortuitously (i.e. neutrally), which may be the starting point for the evolutionary selection of a new function.

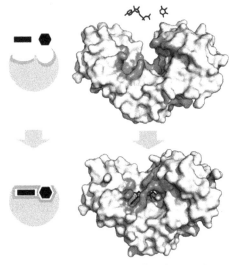

Enzyme changes shape by induced fit upon substrate binding to form enzyme-substrate complex. Hexokinase has a large induced fit motion that closes over the substrates adenosine triphosphate and xylose. Binding sites in blue, substrates in black and Mg^{2+} cofactor in yellow. (PDB 2E2N, 2E2Q)

"Lock and Key" Model

To explain the observed specificity of enzymes, in 1894 Emil Fischer proposed that both the enzyme and the substrate possess specific complementary geometric shapes that fit exactly into one another. This is often referred to as "the lock and key" model. This early model explains enzyme specificity, but fails to explain the stabilization of the transition state that enzymes achieve.

Induced Fit Model

In 1958, Daniel Koshland suggested a modification to the lock and key model since enzymes are rather flexible structures, the active site is continuously reshaped by interactions with the substrate as the substrate interacts with the enzyme. As a result, the substrate does not simply bind to a rigid active site; the amino acid side-chains that make up the active site are molded into the precise positions that enable the enzyme to perform its catalytic function. In some cases, such as glycosidases, the substrate molecule also changes shape slightly as it enters the active site. The active site continues to change until the substrate is completely bound, at which point the final shape and charge distribution is determined. Induced fit may enhance the fidelity of molecular recognition in the presence of competition and noise via the conformational proofreading mechanism.

Catalysis

Enzymes can accelerate reactions in several ways, all of which lower the activation energy ($\Delta G^{\ddagger}$, Gibbs free energy)

1. By stabilizing the transition state

    ○ Creating an environment with a charge distribution complementary to that of the transition state to lower its energy.

2. By providing an alternative reaction pathway

    ○ Temporarily reacting with the substrate, forming a covalent intermediate to provide a lower energy transition state.

3. By destabilising the substrate ground state

    ○ Distorting bound substrate(s) into their transition state form to reduce the energy required to reach the transition state.

    ○ By orienting the substrates into a productive arrangement to reduce the reaction entropy change. The contribution of this mechanism to catalysis is relatively small.

Enzymes may use several of these mechanisms simultaneously. For example, proteases such as trypsin perform covalent catalysis using a catalytic triad, stabilise charge build-up on the transition states using an oxyanion hole, complete hydrolysis using an oriented water substrate.

Dynamics

Enzymes are not rigid, static structures; instead they have complex internal dynamic motions – that is, movements of parts of the enzyme's structure such as individual amino acid residues, groups of residues forming a protein loop or unit of secondary structure, or even an entire protein domain. These motions give rise to a conformational ensemble of slightly different structures that interconvert with one another at equilibrium. Different states within this ensemble may be associated with different aspects of an enzyme's function. For example, different conformations of the enzyme dihydrofolate reductase are associated with the substrate binding, catalysis, cofactor release, and product release steps of the catalytic cycle.

Allosteric Modulation

Allosteric sites are pockets on the enzyme, distinct from the active site, that bind to molecules in the cellular environment. These molecules then cause a change in the conformation or dynamics of the enzyme that is transduced to the active site and thus affects the reaction rate of the enzyme. In this way, allosteric interactions can either inhibit or activate enzymes. Allosteric interactions with metabolites upstream or downstream in an enzyme's metabolic pathway cause feedback regulation, altering the activity of the enzyme according to the flux through the rest of the pathway.

Cofactors

Some enzymes do not need additional components to show full activity. Others require non-protein molecules called cofactors to be bound for activity. Cofactors can be either inorganic (e.g., metal ions and iron-sulfur clusters) or organic compounds (e.g., flavin and heme). Organic cofactors can be either coenzymes, which are released from the enzyme's active site during the reaction, or prosthetic groups, which are tightly bound to an enzyme. Organic prosthetic groups can be covalently bound (e.g., biotin in enzymes such as pyruvate carboxylase).

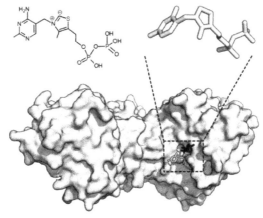

Chemical structure for thiamine pyrophosphate and protein structure of transketolase. Thiamine pyrophosphate cofactor in yellow and xylulose 5-phosphate substrate in black. (PDB 4KXV)

An example of an enzyme that contains a cofactor is carbonic anhydrase, which is shown in the ribbon diagram above with a zinc cofactor bound as part of its active site. These tightly bound ions or molecules are usually found in the active site and are involved in catalysis. For example, flavin and heme cofactors are often involved in redox reactions.

Enzymes that require a cofactor but do not have one bound are called *apoenzymes* or *apoproteins*. An enzyme together with the cofactor(s) required for activity is called a *holoenzyme* (or haloenzyme). The term *holoenzyme* can also be applied to enzymes that contain multiple protein subunits, such as the DNA polymerases; here the holoenzyme is the complete complex containing all the subunits needed for activity.

Coenzymes

Coenzymes are small organic molecules that can be loosely or tightly bound to an enzyme. Coenzymes transport chemical groups from one enzyme to another. Examples include NADH, NADPH and adenosine triphosphate (ATP). Some coenzymes, such as riboflavin, thiamine and folic acid, are vitamins, or compounds that cannot be synthesized by the body and must be acquired from the diet. The chemical groups carried include the hydride ion (H^-) carried by NAD or $NADP^+$, the phosphate group carried by adenosine triphosphate, the acetyl group carried by coenzyme A, formyl, methenyl or methyl groups carried by folic acid and the methyl group carried by S-adenosylmethionine.

Since coenzymes are chemically changed as a consequence of enzyme action, it is useful to consider coenzymes to be a special class of substrates, or second substrates, which are common to many different enzymes. For example, about 1000 enzymes are known to use the coenzyme NADH.

Coenzymes are usually continuously regenerated and their concentrations maintained at a steady level inside the cell. For example, NADPH is regenerated through the pentose phosphate pathway and *S*-adenosylmethionine by methionine adenosyltransferase. This continuous regeneration means that small amounts of coenzymes can be used very intensively. For example, the human body turns over its own weight in ATP each day.

Thermodynamics

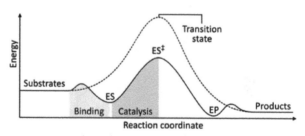

The energies of the stages of a chemical reaction. Uncatalysed (dashed line), substrates need a lot of activation energy to reach a transition state, which then decays into lower-energy products. When enzyme catalysed (solid line), the enzyme binds the substrates (ES), then stabilizes the transition state (ES^*) to reduce the activation energy required to produce products (EP) which are finally released.

As with all catalysts, enzymes do not alter the position of the chemical equilibrium of the reaction. In the presence of an enzyme, the reaction runs in the same direction as it would without the enzyme, just more quickly. For example, carbonic anhydrase catalyzes its reaction in either direction depending on the concentration of its reactants

(in tissues; high CO_2 concentration)   (1) (in lungs; low CO_2 concentration)   (2)

The rate of a reaction is dependent on the activation energy needed to form the transition state which then decays into products. Enzymes increase reaction rates by lowering the energy of the transition state. First, binding forms a low energy enzyme-substrate complex (ES). Secondly the enzyme stabilises the transition state such that it requires less energy to achieve compared to the uncatalyzed reaction ($ES^\ddagger$). Finally the enzyme-product complex (EP) dissociates to release the products.

Enzymes can couple two or more reactions, so that a thermodynamically favorable reaction can be used to "drive" a thermodynamically unfavourable one so that the combined energy of the products is lower than the substrates. For example, the hydrolysis of ATP is often used to drive other chemical reactions.

Kinetics

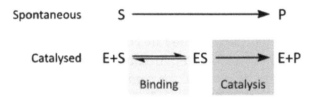

A chemical reaction mechanism with or without enzyme catalysis. The enzyme (E) binds substrate (S) to produce product (P).

Enzyme kinetics is the investigation of how enzymes bind substrates and turn them into products. The rate data used in kinetic analyses are commonly obtained from enzyme assays. In 1913 Leonor Michaelis and Maud Leonora Menten proposed a quantitative theory of enzyme kinetics, which is referred to as Michaelis–Menten kinetics. The major contribution of Michaelis and Menten was to think of enzyme reactions in two stages. In the first, the substrate binds reversibly to the enzyme, forming the enzyme-substrate complex. This is sometimes called the Michaelis-Menten complex in their honor. The enzyme then catalyzes the chemical step in the reaction and releases the product. This work was further developed by G. E. Briggs and J. B. S. Haldane, who derived kinetic equations that are still widely used today.

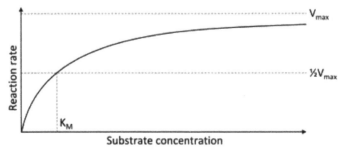

Saturation curve for an enzyme reaction showing the relation between the substrate concentration and reaction rate.

Enzyme rates depend on solution conditions and substrate concentration. To find the maximum speed of an enzymatic reaction, the substrate concentration is increased until a constant rate of product formation is seen. This is shown in the saturation curve on the right. Saturation happens because, as substrate concentration increases, more and more of the free enzyme is converted into the substrate-bound ES complex. At the maximum reaction rate (V_{max}) of the enzyme, all the enzyme active sites are bound to substrate, and the amount of ES complex is the same as the total amount of enzyme.

V_{max} is only one of several important kinetic parameters. The amount of substrate needed to achieve a given rate of reaction is also important. This is given by the Michaelis-Menten constant (K_m), which is the substrate concentration required for an enzyme to reach one-half its maximum reaction rate; generally, each enzyme has a characteristic K_m for a given substrate. Another useful constant is k_{cat}, also called the *turnover number*, which is the number of substrate molecules handled by one active site per second.

The efficiency of an enzyme can be expressed in terms of k_{cat}/K_m. This is also called the specificity constant and incorporates the rate constants for all steps in the reaction up to and including the first irreversible step. Because the specificity constant reflects both affinity and catalytic ability, it is useful for comparing different enzymes against each other, or the same enzyme with different substrates. The theoretical maximum for the specificity constant is called the diffusion limit and is about 10^8 to 10^9 ($M^{-1}\,s^{-1}$). At this point every collision of the enzyme with its substrate will result in catalysis, and the rate of product formation is not limited by the reaction rate but by the diffusion rate. Enzymes with this property are called *catalytically perfect* or *kinetically perfect*. Example of such enzymes are triose-phosphate isomerase, carbonic anhydrase, acetylcholinesterase, catalase, fumarase, β-lactamase, and superoxide dismutase. The turnover of such enzymes can reach several million reactions per second.

Michaelis–Menten kinetics relies on the law of mass action, which is derived from the assumptions of free diffusion and thermodynamically driven random collision. Many biochemical or cellular processes deviate significantly from these conditions, because of macromolecular crowding and constrained molecular movement. More recent, complex extensions of the model attempt to correct for these effects.

Inhibition

The coenzyme folic acid (left) and the anti-cancer drug methotrexate (right) are very similar in structure (differences show in green). As a result, methotrexate is a competitive inhibitor of many enzymes that use folates.

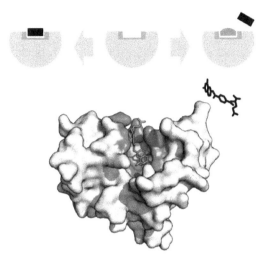

An enzyme binding site that would normally bind substrate can alternatively bind a competitive inhibitor, preventing substrate access. Dihydrofolate reductase is inhibited by methotrexate which prevents binding of its substrate, folic acid. Binding site in blue, inhibitor in green, and substrate in black. (PDB 4QI9)

Enzyme reaction rates can be decreased by various types of enzyme inhibitors.

Types of Inhibition

Competitive

A competitive inhibitor and substrate cannot bind to the enzyme at the same time. Often competitive inhibitors strongly resemble the real substrate of the enzyme. For example, the drug methotrexate is a competitive inhibitor of the enzyme dihydrofolate reductase, which catalyzes the reduction of dihydrofolate to tetrahydrofolate. The similarity between the structures of dihydrofolate and this drug are shown in the accompanying figure. This type of inhibition can be overcome with high substrate concentration. In some cases, the inhibitor can bind to a site other than the binding-site of the usual substrate and exert an allosteric effect to change the shape of the usual binding-site.

Non-competitive

A non-competitive inhibitor binds to a site other than where the substrate binds. The substrate still binds with its usual affinity and hence K_m remains the same. However the inhibitor reduces the catalytic efficiency of the enzyme so that V_{max} is reduced. In contrast to competitive inhibition, non-competitive inhibition cannot be overcome with high substrate concentration.

Uncompetitive

An uncompetitive inhibitor cannot bind to the free enzyme, only to the enzyme-substrate complex; hence, these types of inhibitors are most effective at high substrate concentration. In the presence of the inhibitor, the enzyme-substrate complex is inactive. This type of inhibition is rare.

Mixed

A mixed inhibitor binds to an allosteric site and the binding of the substrate and the inhibitor affect each other. The enzyme's function is reduced but not eliminated when bound to the inhibitor. This type of inhibitor does not follow the Michaelis-Menten equation.

Irreversible

An irreversible inhibitor permanently inactivates the enzyme, usually by forming a covalent bond to the protein. Penicillin and aspirin are common drugs that act in this manner.

Functions of Inhibitors

In many organisms, inhibitors may act as part of a feedback mechanism. If an enzyme produces too much of one substance in the organism, that substance may act as an inhibitor for the enzyme at the beginning of the pathway that produces it, causing production of the substance to slow down or stop when there is sufficient amount. This is a form of negative feedback. Major metabolic pathways such as the citric acid cycle make use of this mechanism.

Since inhibitors modulate the function of enzymes they are often used as drugs. Many such drugs are reversible competitive inhibitors that resemble the enzyme's native substrate, similar to methotrexate above; other well-known examples include statins used to treat high cholesterol, and protease inhibitors used to treat retroviral infections such as HIV. A common example of an irreversible inhibitor that is used as a drug is aspirin, which inhibits the COX-1 and COX-2 enzymes that produce the inflammation messenger prostaglandin. Other enzyme inhibitors are poisons. For example, the poison cyanide is an irreversible enzyme inhibitor that combines with the copper and iron in the active site of the enzyme cytochrome c oxidase and blocks cellular respiration.

Biological Function

Enzymes serve a wide variety of functions inside living organisms. They are indispensable for signal transduction and cell regulation, often via kinases and phosphatases. They also generate movement, with myosin hydrolyzing ATP to generate muscle contraction, and also transport cargo around the cell as part of the cytoskeleton. Other ATPases in the cell membrane are ion pumps involved in active transport. Enzymes are also involved in more exotic functions, such as luciferase generating light in fireflies. Viruses can also contain enzymes for infecting cells, such as the HIV integrase and reverse transcriptase, or for viral release from cells, like the influenza virus neuraminidase.

An important function of enzymes is in the digestive systems of animals. Enzymes such as amylases and proteases break down large molecules (starch or proteins, respectively) into smaller ones, so they can be absorbed by the intestines. Starch molecules, for example, are too large to be absorbed from the intestine, but enzymes hydrolyze the starch chains into smaller molecules such as maltose and eventually glucose, which can then be absorbed. Different enzymes digest different food substances. In ruminants, which have herbivorous diets, microorganisms in the gut produce another enzyme, cellulase, to break down the cellulose cell walls of plant fiber.

Metabolism

Several enzymes can work together in a specific order, creating metabolic pathways. In a metabolic pathway, one enzyme takes the product of another enzyme as a substrate. After the catalytic reaction, the product is then passed on to another enzyme. Sometimes more than one enzyme can catalyze the same reaction in parallel; this can allow more complex regulation with, for example, a low constant activity provided by one enzyme but an inducible high activity from a second enzyme.

The metabolic pathway of glycolysis releases energy by converting glucose to pyruvate by via a series of intermediate metabolites. Each chemical modification (red box) is performed by a different enzyme.

Enzymes determine what steps occur in these pathways. Without enzymes, metabolism would neither progress through the same steps and could not be regulated to serve the needs of the cell. Most central metabolic pathways are regulated at a few key steps, typically through enzymes whose activity involves the hydrolysis of ATP. Because this reaction releases so much energy, other reactions that are thermodynamically unfavorable can be coupled to ATP hydrolysis, driving the overall series of linked metabolic reactions.

Control of Activity

There are five main ways that enzyme activity is controlled in the cell.

Regulation

Enzymes can be either activated or inhibited by other molecules. For example, the end product(s) of a metabolic pathway are often inhibitors for one of the first enzymes of the pathway (usually the first irreversible step, called committed step), thus regulating the amount of end product made by the pathways. Such a regulatory mechanism is called a negative feedback mechanism, because the amount of the end product produced is regulated by its own concentration. Negative feedback mechanism can effectively adjust the rate of synthesis of intermediate metabolites according to the demands of the cells. This helps with effective allocations of materials and energy economy, and it prevents the excess manufacture of end products. Like other homeostatic devices, the control of enzymatic action helps to maintain a stable internal environment in living organisms.

Post-translational modification

Examples of post-translational modification include phosphorylation, myristoylation and glycosylation. For example, in the response to insulin, the phosphorylation of multiple enzymes, including glycogen synthase, helps control the synthesis or degradation of glycogen and allows the cell to respond to changes in blood sugar. Another example of post-translational modification is the cleavage of the polypeptide chain. Chymotrypsin, a digestive protease, is produced in inactive form as chymotrypsinogen in the pancreas and transported in this form to the stomach where it is activated. This stops the enzyme from digesting the pancreas or other tissues before it enters the gut. This type of inactive precursor to an enzyme is known as a zymogen or proenzyme.

Quantity

Enzyme production (transcription and translation of enzyme genes) can be enhanced or diminished by a cell in response to changes in the cell's environment. This form of gene regulation is called enzyme induction. For example, bacteria may become resistant to antibiotics such as penicillin because enzymes called beta-lactamases are induced that hydrolyse the crucial beta-lactam ring within the penicillin molecule. Another example comes from enzymes in the liver called cytochrome P450 oxidases, which are important in drug metabolism. Induction or inhibition of these enzymes can cause drug interactions. Enzyme levels can also be regulated by changing the rate of enzyme degradation.

Subcellular distribution

Enzymes can be compartmentalized, with different metabolic pathways occurring in different cellular compartments. For example, fatty acids are synthesized by one set of enzymes in the cytosol, endoplasmic reticulum and Golgi and used by a different set of enzymes as a source of energy in the mitochondrion, through β-oxidation. In addition, trafficking of the enzyme to different compartments may change the degree of protonation (cytoplasm neutral and lysosome acidic) or oxidative state [e.g., oxidized (periplasm) or reduced (cytoplasm)] which in turn affects enzyme activity.

Organ specialization

In multicellular eukaryotes, cells in different organs and tissues have different patterns of gene expression and therefore have different sets of enzymes (known as isozymes) available for metabolic reactions. This provides a mechanism for regulating the overall metabolism of the organism. For example, hexokinase, the first enzyme in the glycolysis pathway, has a specialized form called glucokinase expressed in the liver and pancreas that has a lower affinity for glucose yet is more sensitive to glucose concentration. This enzyme is involved in sensing blood sugar and regulating insulin production.

Involvement in Disease

Since the tight control of enzyme activity is essential for homeostasis, any malfunction (mutation, overproduction, underproduction or deletion) of a single critical enzyme can lead to a genetic disease. The malfunction of just one type of enzyme out of the thousands of types present in the human body can be fatal. An example of a fatal genetic disease due to enzyme insufficiency is Tay-Sachs disease, in which patients lack the enzyme hexosaminidase.

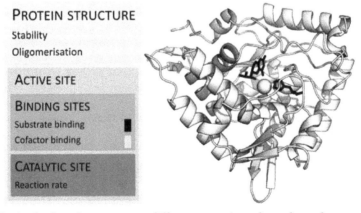

In phenylalanine hydroxylase over 300 different mutations throughout the structure cause phenylketonuria. Phenylalanine substrate and tetrahydrobiopterin coenzyme in black, and Fe^{2+} cofactor in yellow. (PDB 1KW0)

One example of enzyme deficiency is the most common type of phenylketonuria. Many different single amino acid mutations in the enzyme phenylalanine hydroxylase, which catalyzes the first step in the degradation of phenylalanine, result in build-up of phenylalanine and related products. Some mutations are in the active site, directly disrupting binding and catalysis, but many are far from the active site and reduce activity by destabilising the protein structure, or affecting correct oligomerisation. This can lead to intellectual disability if the disease is untreated. Another example is pseudocholinesterase deficiency, in which the body's ability to break down choline ester drugs is impaired. Oral administration of enzymes can be used to treat some functional enzyme deficiencies, such as pancreatic insufficiency and lactose intolerance.

Another way enzyme malfunctions can cause disease comes from germline mutations in genes coding for DNA repair enzymes. Defects in these enzymes cause cancer because cells are less able to repair mutations in their genomes. This causes a slow accumulation of mutations and results in the development of cancers. An example of such a hereditary cancer syndrome is xeroderma pigmentosum, which causes the development of skin cancers in response to even minimal exposure to ultraviolet light.

Naming Conventions

An enzyme's name is often derived from its substrate or the chemical reaction it catalyzes, with the word ending in -*ase*. Examples are lactase, alcohol dehydrogenase and DNA polymerase. Different enzymes that catalyze the same chemical reaction are called isozymes.

The International Union of Biochemistry and Molecular Biology have developed a nomenclature for enzymes, the EC numbers; each enzyme is described by a sequence of four numbers preceded by "EC". The first number broadly classifies the enzyme based on its mechanism.

The top-level classification is

- EC 1, Oxidoreductases catalyze oxidation/reduction reactions
- EC 2, Transferases transfer a functional group (*e.g.* a methyl or phosphate group)
- EC 3, Hydrolases catalyze the hydrolysis of various bonds
- EC 4, Lyases cleave various bonds by means other than hydrolysis and oxidation
- EC 5, Isomerases catalyze isomerization changes within a single molecule
- EC 6, Ligases join two molecules with covalent bonds.

These sections are subdivided by other features such as the substrate, products, and chemical mechanism. An enzyme is fully specified by four numerical designations. For example, hexokinase (EC 2.7.1.1) is a transferase (EC 2) that adds a phosphate group (EC 2.7) to a hexose sugar, a molecule containing an alcohol group (EC 2.7.1).

Industrial Applications

Enzymes are used in the chemical industry and other industrial applications when extremely specific catalysts are required. Enzymes in general are limited in the number of reactions they have evolved to catalyze and also by their lack of stability in organic solvents and at high temperatures. As a consequence, protein engineering is an active area of research and involves attempts to create new enzymes with novel properties, either through rational design or *in vitro* evolution. These efforts have begun to be successful, and a few enzymes have now been designed "from scratch" to catalyze reactions that do not occur in nature.

Application	Enzymes used	Uses
Biofuel industry	Cellulases	Break down cellulose into sugars that can be fermented to produce cellulosic ethanol.
	Ligninases	Pretreatment of biomass for biofuel production.
Biological detergent	Proteases, amylases, lipases	Remove protein, starch, and fat or oil stains from laundry and dishware.
	Mannanases	Remove food stains from the common food additive guar gum.
Brewing industry	Amylase, glucanases, proteases	Split polysaccharides and proteins in the malt.
	Betaglucanases	Improve the wort and beer filtration characteristics.
	Amyloglucosidase and pullulanases	Make low-calorie beer and adjust fermentability.
	Acetolactate decarboxylase (ALDC)	Increase fermentation efficiency by reducing diacetyl formation.
Culinary uses	Papain	Tenderize meat for cooking.
Dairy industry	Rennin	Hydrolyze protein in the manufacture of cheese.
	Lipases	Produce Camembert cheese and blue cheeses such as Roquefort.

Food processing	Amylases	Produce sugars from starch, such as in making high-fructose corn syrup.
	Proteases	Lower the protein level of flour, as in biscuit-making.
	Trypsin	Manufacture hypoallergenic baby foods.
	Cellulases, pectinases	Clarify fruit juices.
Molecular biology	Nucleases, DNA ligase and polymerases	Use restriction digestion and the polymerase chain reaction to create recombinant DNA.
Paper industry	Xylanases, hemicellulases and lignin peroxidases	Remove lignin from kraft pulp.
Personal care	Proteases	Remove proteins on contact lenses to prevent infections.
Starch industry	Amylases	Convert starch into glucose and various syrups.

References

- Petsko GA, Ringe D (2003). "Chapter 1: From sequence to structure". Protein structure and function. London: New Science. p. 27. ISBN 978-1405119221.

- Suzuki H (2015). "Chapter 7: Active Site Structure". How Enzymes Work: From Structure to Function. Boca Raton, FL: CRC Press. pp. 117–140. ISBN 978-981-4463-92-8.

- Krauss G (2003). "The Regulations of Enzyme Activity". Biochemistry of Signal Transduction and Regulation (3rd ed.). Weinheim: Wiley-VCH. pp. 89–114. ISBN 9783527605767.

- Cooper GM (2000). "Chapter 2.2: The Central Role of Enzymes as Biological Catalysts". The Cell: a Molecular Approach (2nd ed.). Washington (DC): ASM Press. ISBN 0-87893-106-6.

- Cox MM, Nelson DL (2013). "Chapter 6.2: How enzymes work". Lehninger Principles of Biochemistry (6th ed.). New York, N.Y.: W.H. Freeman. p. 195. ISBN 978-1464109621.

- McArdle WD, Katch F, Katch VL (2006). "Chapter 9: The Pulmonary System and Exercise". Essentials of Exercise Physiology (3rd ed.). Baltimore, Maryland: Lippincott Williams & Wilkins. pp. 312–3. ISBN 978-0781749916.

- Suzuki H (2015). "Chapter 8: Control of Enzyme Activity". How Enzymes Work: From Structure to Function. Boca Raton, FL: CRC Press. pp. 141–69. ISBN 978-981-4463-92-8.

- Suzuki H (2015). "Chapter 4: Effect of pH, Temperature, and High Pressure on Enzymatic Activity". How Enzymes Work: From Structure to Function. Boca Raton, FL: CRC Press. pp. 53–74. ISBN 978-981-4463-92-8.

- James WD, Elston D, Berger TG (2011). Andrews' Diseases of the Skin: Clinical Dermatology (11th ed.). London: Saunders/ Elsevier. p. 567. ISBN 978-1437703146.

- Tarté R (2008). Ingredients in Meat Products Properties, Functionality and Applications. New York: Springer. p. 177. ISBN 978-0-387-71327-4.

- Farris PL (2009). "Economic Growth and Organization of the U.S. Starch Industry". In BeMiller JN, Whistler RL. Starch Chemistry and Technology (3rd ed.). London: Academic. ISBN 9780080926551.

Types of Enzyme

Enzymes are classified according to the way they work on a molecular level into categories like- transferases, isomerases etc. According to the site of their activity they can be divided into two categories- endoenzyme and exoenzyme. This chapter studies the categories comprehensively with suitable examples to help the reader understand the intricacies of each category.

Transferase

A transferase is any one of a class of enzymes that enact the transfer of specififunctional groups (e.g. a methyl or glycosyl group) from one molecule (called the donor) to another (called the acceptor). They are involved in hundreds of different biochemical pathways throughout biology, and are integral to some of life's most important processes.

RNA polymerase from *Saccharomyces cerevisiae* complexed with α-amanitin (in red). Despite the use of the term "polymerase," RNA polymerases are classified as a form of nucleotidyl transferase.

Transferases are involved in myriad reactions in the cell. Some examples of these reactions include the activity of coenzyme A (CoA) transferase, which transfers thiol esters, the action of N-acetyltransferase is part of the pathway that metabolizes tryptophan, and also includes the regulation of pyruvate dehydrogenase (PDH), which converts pyruvate to acetyl CoA. Transferases are also utilized during translation. In this case, an amino acid chain is the functional group transferred by a peptidyl transferase. The transfer involves the removal of the growing amino acid chain from the tRNA molecule in the A-site of the ribosome and its subsequent addition to the amino acid attached to the tRNA in the P-site.

Mechanistically, an enzyme that catalyzed the following reaction would be a transferase:

$$Xgroup + Y \xrightarrow[\text{transferase}]{} X + Ygroup$$

In the above reaction, X would be the donor, and Y would be the acceptor. "Group" would be the functional group transferred as a result of transferase activity. The donor is often a coenzyme.

History

Some of the most important discoveries relating to transferases occurred as early as the 1930s. Earliest discoveries of transferase activity occurred in other classifications of enzymes, including Beta-galactosidase, protease, and acid/base phosphatase. Prior to the realization that individual enzymes were capable of such a task, it was believed that two or more enzymes enacted functional group transfers.

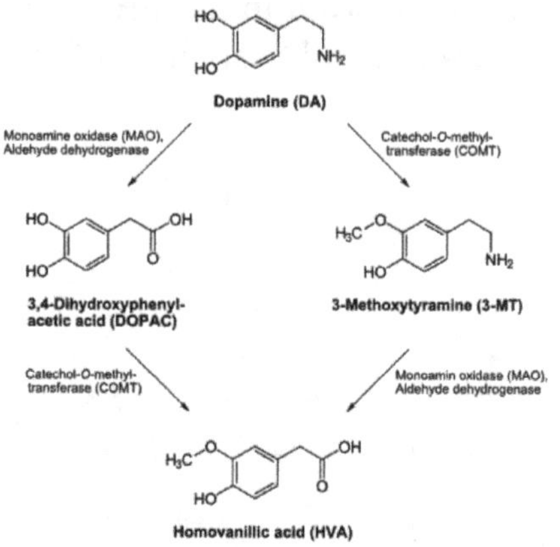

Biodegradation of dopamine via catechol-O-methyltransferase (along with other enzymes). The mechanism for dopamine degradation led to the Nobel Prize in Physiology or Medicine in 1970.

Transamination, or the transfer of an amine (or NH_2) group from an amino acid to a keto acid by an aminotransferase (also known as a "transaminase"), was first noted in 1930 by D. M. Needham, after observing the disappearance of glutamiacid added to pigeon breast muscle. This observance was later verified by the discovery of its reaction mechanism by Braunstein and Kritzmann in 1937. Their analysis showed that this reversible reaction could be applied to other tissues. This assertion was validated by Rudolf Schoenheimer's work with radioisotopes as tracers in 1937. This in turn would pave the way for the possibility that similar transfers were a primary means of producing most amino acids via amino transfer.

Another such example of early transferase research and later reclassification involved the discovery of uridyl transferase. In 1953, the enzyme UDP-glucose pyrophosphory-lase was shown to be a transferase, when it was found that it could reversibly produce UTP and G1P from UDP-glucose and an organipyrophosphate.

Another example of historical significance relating to transferase is the discovery of the mechanism of catecholamine breakdown by catechol-O-methyltransferase. This dis-covery was a large part of the reason for Julius Axelrod's 1970 Nobel Prize in Physiology or Medicine (shared with Sir Bernard Katz and Ulf von Euler).

Classification of transferases continues to this day, with new ones being discovered frequently. An example of this is Pipe, a sulfotransferase involved in the dorsal-ventral patterning of *Drosophilia*. Initially, the exact mechanism of Pipe was unknown, due to a lack of information on its substrate. Research into Pipe's catalytiactivity eliminated the likelihood of it being a heparan sulfate glycosaminoglycan. Further research has shown that Pipe targets the ovarian structures for sulfation. Pipe is currently classified as a *Drosophilia* heparan sulfate 2-O-sulfotransferase.

Nomenclature

Systematinames of transferases are constructed in the form of "donor:acceptor grouptransferase." For example, methylamine:L-glutamate N-methyltransferase would be the standard naming convention for the transferase methylamine-glutamate N-methyltransferase, where methylamine is the donor, L-glutamate is the acceptor, and methyltransferase is the Ecategory grouping. This same action by the transferase can be illustrated as follows:

$$\text{methylamine} + \text{L-glutamate} \rightleftharpoons NH_3 + \text{N-methyl-L-glutamate}$$

However, other accepted names are more frequently used for transferases, and are of-ten formed as "acceptor grouptransferase" or "donor grouptransferase." For example, a DNA methyltransferase is a transferase that catalyzes the transfer of a methyl group to a DNA acceptor. In practice, many molecules are not referred to using this termi-nology due to more prevalent common names. For example, RNA Polymerase is the modern common name for what was formerly known as RNA nucleotidyltransferase, a kind of nucleotidyl transferase that transfers nucleotides to the 3' end of a growing RNA strand. In the Esystem of classification, the accepted name for RNA Polymerase is DNA-directed RNA polymerase.

Classification

Described primarily based on the type of biochemical group transferred, transferases can be divided into ten categories (based on the ENumber classification). These cate-gories comprise over 450 different unique enzymes. In the Enumbering system, trans-ferases have been given a classification of EC2. Hydrogen is not considered a functional group when it comes to transferase targets; instead, hydrogen transfer is included under oxidoreductases, due to electron transfer considerations.

Classification of transferases into subclasses		
Enumber	Examples	Group(s) transferred
E2.1	methyltransferase and formyltransferase	single-carbon groups
E2.2	transketolase and transaldolase	aldehyde or ketone groups
E2.3	acyltransferase	acyl groups or groups that become alkyl groups during transfer
E2.4	glycosyltransferase, hexosyltransferase, and pentosyltransferase	glycosyl groups, as well as hexoses and pentoses
E2.5	riboflavin synthase and chlorophyll synthase	alkyl or aryl groups, other than methyl groups
E2.6	transaminase, and oximinotransferase	nitrogenous groups
E2.7	phosphotransferase, polymerase, and kinase	phosphorus-containing groups; subclasses are based on the acceptor (e.g. alcohol, carboxyl, etc.)
E2.8	sulfurtransferase and sulfotransferase	sulfur-containing groups
E2.9	selenotransferase	selenium-containing groups
E2.10	molybdenumtransferase and tungstentransferase	molybdenum or tungsten

Reactions

E2.1: Single Carbon Transferases

E2.1 includes enzymes that transfer single-carbon groups. This category consists of transfers of methyl, hydroxymethyl, formyl, carboxy, carbamoyl, and amido groups. Carbamoyltransferases, as an example, transfer a carbamoyl group from one molecule to another. Carbamoyl groups follow the formula NH_2CO. In ATCase such a transfer is written as Carbamyl phosphate + L-aspertate $\rightarrow$ L-carbamyl aspartate + phosphate.

Reaction involving aspartate transcarbamylase.

E2.2: Aldehyde and Ketone Transferases

Enzymes that transfer aldehyde or ketone groups and included in E2.2. This category consists of various transketolases and transaldolases. Transaldolase, the namesake of aldehyde transferases, is an important part of the pentose phosphate pathway. The reaction it catalyzes consists of a transfer of a dihydroxyacetone functional group to Glyceraldehyde 3-phosphate (also known as G3P). The reaction is as follows: sedoheptulose 7-phosphate + glyceraldehyde 3-phosphate $\rightleftharpoons$ erythrose 4-phosphate + fructose 6-phosphate.

The reaction catalyzed by transaldolase

E2.3: Acyl Transferases

Transfer of acyl groups or acyl groups that become alkyl groups during the process of being transferred are key aspects of E2.3. Further, this category also differentiates between amino-acyl and non-amino-acyl groups. Peptidyl transferase is a ribozyme that facilitates formation of peptide bonds during translation. As an aminoacyltransferase, it catalyzes the transfer of a peptide to an aminoacyl-tRNA, following this reaction: peptidyl-tRNA$_A$ + aminoacyl-tRNA$_B$ $\rightleftharpoons$ $\rightleftharpoons$ tRNA$_A$ + peptidyl aminoacyl-tRNA$_B$.

E2.4: Glycosyl, Hexosyl, and Pentosyl Transferases

E2.4 includes enzymes that transfer glycosyl groups, as well as those that transfer hexose and pentose. Glycosyltransferase is a subcategory of E2.4 transferases that is involved in biosynthesis of disaccharides and polysaccharides through transfer of monosaccharides to other molecules. An example of a prominent glycosyltransferase is lactose synthase which is a dimer possessing two protein subunits. Its primary action is to produce lactose from glucose and UDP-galactose. This occurs via the following pathway: UDP-β-D-galactose + D-glucose $\rightleftharpoons$ UDP + lactose.

E2.5: Alkyl and Aryl Transferases

E2.5 relates to enzymes that transfer alkyl or aryl groups, but does not include methyl groups. This is in contrast to functional groups that become alkyl groups when transferred, as those are included in E2.3. E2.5 currently only possesses one sub-class: Alkyl and aryl transferases. Cysteine synthase, for example, catalyzes the formation of acetiacids and cysteine from O_3-acetyl-L-serine and hydrogen sulfide: O_3-acetyl-L-serine + H_2S $\rightleftharpoons$ L-cysteine + acetate.

E2.6: Nitrogenous Transferases

The grouping consistent with transfer of nitrogenous groups is E2.6. This includes enzymes like transaminase (also known as "aminotransferase"), and a very small number of oximinotransferases and other nitrogen group transferring enzymes. E2.6 previously included amidinotransferase but it has since been reclassified as a subcategory of E2.1 (single-carbon transferring enzymes). In the case of aspartate transaminase, which can act on tyrosine, phenylalanine, and tryptophan, it reversibly transfers an amino group from one molecule to the other.

Aspartate aminotransferase can act on several different amino acids

The reaction, for example, follows the following order: L-aspartate +2-oxoglutarate ⇌ oxaloacetate + L-glutamate.

E2.7: Phosphorus Transferases

While E2.7 includes enzymes that transfer phosphorus-containing groups, it also includes nuclotidyl transferases as well. Sub-category phosphotransferase is divided up in categories based on the type of group that accepts the transfer. Groups that are classified as phosphate acceptors include: alcohols, carboxy groups, nitrogenous groups, and phosphate groups. Further constituents of this subclass of transferases are various kinases. A prominent kinase is cyclin-dependent kinase (or CDK), which comprises a sub-family of protein kinases. As their name implies, CDKs are heavily dependent on specificyclin molecules for activation. Once combined, the CDK-cyclin complex is capable of enacting its function within the cell cycle.

The reaction catalyzed by CDK is as follows: ATP + a target protein →ADP + a phosphoprotein.

E2.8: Sulfur Transferases

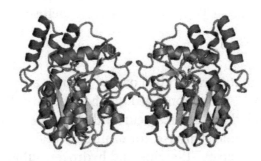

Ribbon diagram of a variant structure of estrogen sulfotransferase (PDB 1aqy EBI)

Transfer of sulfur-containing groups is covered by E2.8 and is subdivided into the sub-categories of sulfurtransferases, sulfotransferases, and CoA-transferases, as well as enzymes that transfer alkylthio groups. A specifigroup of sulfotransferases are those that use PAPS as a sulfate group donor. Within this group is alcohol sulfotransferase which has a broad targeting capacity. Due to this, alcohol sulfotransferase is also known by several other names including "hydroxysteroid sulfotransferase," "steroid sulfokinase," and "estrogen sulfotransferase." Decreases in its activity has been linked to human liver disease. This transferase acts via the following reaction: 3'-phosphoadenylyl sulfate + an alcohol $\rightleftharpoons$ adenosine 3',5'bisphosphate + an alkyl sulfate.

E2.9: Selenium Transferases

E2.9 includes enzymes that transfer selenium-containing groups. This category only contains two transferases, and thus is one of the smallest categories of transferase. Selenocysteine synthase, which was first added to the classification system in 1999, converts seryl-tRNA(SeUCA) into selenocysteyl-tRNA(SeUCA).

E2.10: Metal Transferases

The category of E2.10 includes enzymes that transfer molybdenum or tungsten-containing groups. However, as of 2011, only one enzyme has been added: molybdopterin molybdotransferase. This enzyme is a component of MoCo biosynthesis in *Escherichia coli*. The reaction it catalyzes is as follows: adenylyl-molybdopterin + molybdate $\rightarrow$ molybdenum cofactor $+$ AMP.

Role in Histo-blood Group

The A and B transferases are the foundation of the human ABO blood group system. Both A and B transferases are glycosyltransferases, meaning they transfer a sugar molecule onto an H-antigen. This allows H-antigen to synthesize the glycoprotein and glycolipid conjugates that are known as the A/B antigens. The full name of A transferase is alpha 1-3-N-acetylgalactosaminyltransferase and its function in the cell is to add N-acetylgalactosamine to H-antigen, creating A-antigen. The full name of B transferase is alpha 1-3-galactosyltransferase, and its function in the cell is to add a galactose molecule to H-antigen, creating B-antigen.

It is possible for *Homo sapiens* to have any of four different blood types: Type A (express A antigens), Type B (express B antigens), Type AB (express both A and B antigens) and Type O (express neither A nor B antigens). The gene for A and B transferases is located on chromosome nine. The gene contains seven exons and six introns and the gene itself is over 18kb long. The alleles for A and B transferases are extremely similar. The resulting enzymes only differ in 4 amino acid residues. The differing residues are located at positions 176, 235, 266, and 268 in the enzymes.

Deficiencies

Transferase deficiencies are at the root of many common illnesses. The most common result of a transferase deficiency is a buildup of a cellular product.

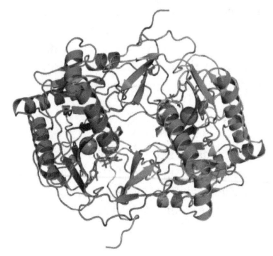

A deficiency of this transferase, E. coli galactose-1-phosphate uridyltransferase is a known cause of galactosemia

SCOT Deficiency

Succinyl-CoA:3-ketoacid CoA transferase deficiency (or SCOT deficiency) leads to a buildup of ketones. Ketones are created upon the breakdown of fats in the body and are an important energy source. Inability to utilize ketones leads to intermittent ketoacidosis, which usually first manifests during infancy. Disease sufferers experience nausea, vomiting, inability to feed, and breathing difficulties. In extreme cases, ketoacidosis can lead to coma and death. The deficiency is caused by mutation in the gene OXTC1. Treatments mostly rely on controlling the diet of the patient.

CPT-II Deficiency

Carnitine palmitoyltransferase II deficiency (also known as CPT-II deficiency) leads to an excess long chain fatty acids, as the body lacks the ability to transport fatty acids into the mitochondria to be processed as a fuel source. The disease is caused by a defect in the gene CPT2. This deficiency will present in patients in one of three ways: lethal neonatal, severe infantile hepatocardiomuscular, and myopathiform. The myopathii is the least severe form of the deficiency and can manifest at any point in the lifespan of the patient. The other two forms appear in infancy. Common symptoms of the lethal neonatal form and the severe infantile forms are liver failure, heart problems, seizures and death. The myopathiform is characterized by muscle pain and weakness following vigorous exercise. Treatment generally includes dietary modifications and carnitine supplements.

Galactosemia

Galactosemia results from an inability to process galactose, a simple sugar. This deficiency occurs when the gene for galactose-1-phosphate uridylyltransferase (GALT) has any number of mutations, leading to a deficiency in the amount of GALT produced. There are two forms of Galactosemia: classiand Duarte. Duarte galactosemia is generally less severe than classigalactosemia and is caused by a deficiency of galactokinase. Galactosemia renders infants unable to process the sugars in breast milk, which leads to vomiting and anorexia within days of birth. Most symptoms of the disease are caused by a buildup of galactose-1-phosphate in the body. Common symptoms include liver failure, sepsis, failure to grow, and mental impairment, among others. Buildup of a second toxisubstance, galactitol, occurs in the lenses of the eyes, causing cataracts. Currently, the only available treatment is early diagnosis followed by adherence to a diet devoid of lactose, and prescription of antibiotics for infections that may develop.

Choline Acetyltransferase Deficiencies

Choline acetyltransferase (also known as ChAT or CAT) is an important enzyme which produces the neurotransmitter acetylcholine. Acetylcholine is involved in many neuropsychifunctions such as memory, attention, sleep and arousal. The enzyme is globular in shape and consists of a single amino acid chain. ChAT functions to transfer an acetyl group from acetyl co-enzyme A to choline in the synapses of nerve cells and exists in two forms: soluble and membrane bound. The ChAT gene is located on chromosome 10.

Alzheimer's Disease

Decreased expression of ChAT is one of the hallmarks of Alzheimer's disease. Patients with Alzheimer's disease show a 30 to 90% reduction in activity in several regions of the brain, including the temporal lobe, the parietal lobe and the frontal lobe. However, ChAT deficiency is not believed to be the main cause of this disease.

Amyotrophilateral Sclerosis (ALS or Lou Gehrig's Disease)

Patients with ALS show a marked decrease in ChAT activity in motor neurons in the spinal cord and brain. Low levels of ChAT activity are an early indication of the disease and are detectable long before motor neurons begin to die. This can even be detected before the patient is symptomatic.

Huntington's Disease

Patients with Huntington's also show a marked decrease in ChAT production. Though the specificause of the reduced production is not clear, it is believed that the death of medium-sized motor neurons with spiny dendrites leads to the lower levels of ChAT production.

Schizophrenia

Patients with Schizophrenia also exhibit decreased levels of ChAT, localized to the mesopontine tegment of the brain and the nucleus accumbens, which is believed to correlate with the decreased cognitive functioning experienced by these patients.

Sudden Infant Death Syndrome (SIDS)

Recent studies have shown that SIDS infants show decreased levels of ChAT in both the hypothalamus and the striatum. SIDS infants also display fewer neurons capable of producing ChAT in the vagus system. These defects in the medulla could lead to an inability to control essential autonomifunctions such as the cardiovascular and respiratory systems.

Congenital Myasthenisyndrome (CMS)

CMS is a family of diseases that are characterized by defects in neuromuscular transmission which leads to recurrent bouts of apnea (inability to breathe) that can be fatal. ChAT deficiency is implicated in myasthenia syndromes where the transition problem occurs presynaptically. These syndromes are characterized by the patients' inability to resynthesize acetylcholine.

Uses in Biotechnology

Terminal Transferases

Terminal transferases are transferases that can be used to label DNA or to produce plasmid vectors. It accomplishes both of these tasks by adding deoxynucleotides in the form of a template to the downstream end or 3' end of an existing DNA molecule. Terminal transferase is one of the few DNA polymerases that can function without an RNA primer.

Glutathione Transferases

The family of glutathione transferases (GST) is extremely diverse, and therefore can be used for a number of biotechnological purposes. Plants use glutathione transferases as a means to segregate toximetals from the rest of the cell. These glutathione transferases can be used to create biosensors to detect contaminants such as herbicides and insecticides. Glutathione transferases are also used in transgeniplants to increase resistance to both biotiand abiotistress. Glutathione transferases are currently being explored as targets for anti-cancer medications due to their role in drug resistance. Further, glutathione transferase genes have been investigated due to their ability to prevent oxidative damage and have shown improved resistance in transgenicultigens.

Rubber Transferases

Currently the only available commercial source of natural rubber is the Hevea plant (Hevea brasiliensis). Natural rubber is superior to synthetirubber in a number of commercial uses. Efforts are being made to produce transgeniplants capable of synthesizing natural rubber, including tobacco and sunflower. These efforts are focused on sequencing the subunits of the rubber transferase enzyme complex in order to transfect these genes into other plants.

Isomerase

Isomerases are a general class of enzymes which convert a molecule from one isomer to another. Isomerases can either facilitate intramolecular rearrangements in which bonds are broken and formed or they can catalyze conformational changes. The general form of such a reaction is as follows:

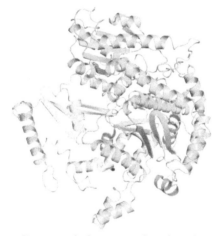

Ribbon diagram of Glucose-6-Phosphate isomerase

$$A\text{–}B \rightarrow B\text{–}A$$

There is only one substrate yielding one product. This product has the same molecular formula as the substrate but differs in bond connectivity or spatial arrangements. Isomerases catalyze reactions across many biological processes, such as in glycolysis and carbohydrate metabolism.

Isomerization

Examples of Isomers

Isomerases catalyze changes within one molecule. They convert one isomer to another, meaning that the end product has the same molecular formula but a different physical structure. Isomers themselves exist in many varieties but can generally be classified as structural isomers or stereoisomers. Structural isomers have a different ordering of bonds and/or different bond connectivity from one another, as in the case of hexane and its four other isomeriforms (2-methylpentane, 3-methylpentane, 2,2-dimethylbutane, and 2,3-dimethylbutane).

The structural isomers of hexane

Stereoisomers have the same ordering of individual bonds and the same connectivity but the three-dimensional arrangement of bonded atoms differ. For example, 2-butene exists in two isomeriforms: *cis*-2-butene and *trans*-2-butene. The sub-categories of isomerases containing racemases, epimerases and cis-trans isomers are examples of enzymes catalyzing the interconversion of stereoisomers. Intramolecular lyases, oxidoreductases and transferases catalyze the interconversion of structural isomers.

Cis-2-butene and Trans-2-butene

The prevalence of each isomer in nature depends in part on the isomerization energy, the difference in energy between isomers. Isomers close in energy can interconvert easily and are often seen in comparable proportions. The isomerization energy, for example, for converting from a stable *cis* isomer to the less stable *trans* isomer is greater than for the reverse reaction, explaining why in the absence of isomerases or an outside energy source such as ultraviolet radiation a given *cis* isomer tends to be present in greater amounts than the *trans* isomer. Isomerases can increase the reaction rate by lowering the isomerization energy.

Calculating isomerase kinetics from experimental data can be more difficult than for other enzymes because the use of product inhibition experiments is impractical. That is, isomerization is not an irreversible reaction since a reaction vessel will contain one substrate and one product so the typical simplified model for calculating reaction kinetics does not hold. There are also practical difficulties in determining the rate-determining step at high concentrations in a single isomerization. Instead, tracer perturbation can overcome these technical difficulties if there are two forms of the unbound enzyme. This technique uses isotope exchange to measure indirectly the interconversion of the free enzyme between its two forms. The radiolabeled substrate and product diffuse in a time-dependent manner. When the system reaches equilibrium the addition of unlabeled substrate perturbs or unbalances it. As equilibrium is established again, the radiolabeled substrate and product are tracked to determine energetiinformation.

```
        CHO                    CHO
         |                      |
    H—C—OH                 HO—C—H
         |                      |
   HO—C—H                  HO—C—H
         |                      |
    H—C—OH                  H—C—OH
         |                      |
    H—C—OH                  H—C—OH
         |                      |
       CH₂OH                  CH₂OH

     D-Glucose              D-Mannose
```

Epimers: D-glucose and D-mannose

The earliest use of this technique elucidated the kinetics and mechanism underlying the action of phosphoglucomutase, favoring the model of indirect transfer of phosphate with one intermediate and the direct transfer of glucose. This technique was then adopted to study the profile of proline racemase and its two states: the form which isomerizes L-proline and the other for D-proline. At high concentrations it was shown that the transition state in this interconversion is rate-limiting and that these enzyme forms may differ just in the protonation at the acidiand basigroups of the active site.

Nomenclature

Generally, "the names of isomerases are formed as "*substrate* isomerase" (for example, enoyl CoA isomerase), or as "*substrate type of isomerase*" (for example, phosphoglucomutase)."

Classification

Enzyme-catalyzed reactions each have a uniquely assigned classification number. Isomerase-catalyzed reactions have their own Ecategory: E5. Isomerases are further classified into six subclasses:

Racemases, Epimerases

This category (E5.1) includes (racemases) and epimerases). These isomerases invert stereochemistry at the target chiral carbon. Racemases act upon molecules with one chiral carbon for inversion of stereochemistry, whereas epimerases target molecules with multiple chiral carbons and act upon one of them. A molecule with only one chiral carbon has two enantiomeriforms, such as serine having the isoforms D-serine and L-serine differing only in the absolute configuration about the chiral carbon. A molecule with multiple chiral carbons has two forms at each chiral carbon. Isomerization at one chiral carbon of several yields epimers, which differ from one another in absolute configuration at just one chiral carbon. For example, D-glucose and D-mannose differ in configuration at just one chiral carbon. This class is further broken down by the group the enzyme acts upon:

Racemases and epimerases:		
Enumber	Description	Examples
E5.1.1	Acting on Amino Acids and Derivative	alanine racemase, methionine racemase
E5.1.2	Acting on Hydroxy Acids and Derivatives	lactate racemase, tartrate epimerase
E5.1.3	Acting on Carbohydrates and Derivatives	ribulose-phosphate 3-epimerase, UDP-glucose 4-epimerase
E5.1.99	Acting on Other Compounds	methylmalonyl CoA epimerase, hydantoin racemase

Cis-trans Isomerases

This category (E5.2) includes enzymes that catalyze the isomerization of cis-trans isomers. Alkenes and cycloalkanes may have cis-trans stereoisomers. These isomers are not distinguished by absolute configuration but rather by the position of substituent groups relative to a plane of reference, as across a double bond or relative to a ring structure. *Cis* isomers have substituent groups on the same side and *trans* isomers have groups on opposite sides.

This category is not broken down any further. All entries presently include:

conversion mediated by peptidylprolyl isomerase

Cis-trans isomerases:	
Enumber	Examples
E5.2.1.1	Maleate isomerase
E5.2.1.2	Maleylacetoacetate isomerase
E5.2.1.4	Maleylpyruvate isomerase
E5.2.1.5	Linoleate isomerase
E5.2.1.6	Furylfuramide isomerase
E5.2.1.8	Peptidylprolyl isomerase
E5.2.1.9	Farnesol 2-isomerase
E5.2.1.10	2-chloro-4-carboxymethylenebut-2-en-1,4-olide isomerase
E5.2.1.12	Zeta-carotene isomerase
E5.2.1.13	Prolycopene isomerase
E5.2.1.14	Beta-carotene isomerase

Intramolecular Oxidoreductases

This category (E5.3) includes intramolecular oxidoreductases. These isomerases catalyze the transfer of electrons from one part of the molecule to another. In other words, they catalyze the oxidation of one part of the molecule and the concurrent reduction of another part. Sub-categories of this class are:

reaction catalyzed by phosphoribosylanthranilate isomerase

Intramolecular oxidoreductases:		
Enumber	Description	Examples
E5.3.1	Interconverting Aldoses and Ketoses	Triose-phosphate isomerase, Ribose-5-phosphate isomerase
E5.3.2	Interconverting Keto- and Enol-Groups	Phenylpyruvate tautomerase, Oxaloacetate tautomerase
E5.3.3	Transposing C=Double Bonds	Steroid Delta-isomerase, L-dopachrome isomerase

| E5.3.4 | Transposing S-S Bonds | Protein disulfide-isomerase |
| E5.3.99 | Other Intramolecular Oxidoreductases | Prostaglandin-D synthase, Allene-oxide cyclase |

Intramolecular Transferases

This category (E5.4) includes intramolecular transferases (mutases). These isomerases catalyze the transfer of functional groups from one part of a molecule to another. Phosphotransferases (E5.4.2) were categorized as transferases (E2.7.5) with regeneration of donors until 1983. This sub-class can be broken down according to the functional group the enzyme transfers:

reaction catalyzed by phosphoenolpyruvate mutase

Intramolecular transferases:		
Enumber	Description	Examples
E5.4.1	Transferring Acyl Groups	Lysolecithin acylmutase, Precorrin-8X methylmutase
E5.4.2	Phosphotransferases (Phosphomutases)	Phosphoglucomutase, Phosphopentomutase
E5.4.3	Transferring Amino Groups	Beta-lysine 5,6-aminomutase, Tyrosine 2,3-aminomutase
E5.4.4	Transferring hydroxy groups	(hydroxyamino)benzene mutase, Isochorismate synthase
E5.4.99	Transferring Other Groups	Methylaspartate mutase, Chorismate mutase

Intramolecular Lyases

This category (E5.5) includes intramolecular lyases. These enzymes catalyze "reactions in which a group can be regarded as eliminated from one part of a molecule, leaving a double bond, while remaining covalently attached to the molecule." Some of these catalyzed reactions involve the breaking of a ring structure.

This category is not broken down any further. All entries presently include:

reaction catalyzed by ent-Copalyl diphosphate synthase

Intramolecular lyases:	
Enumber	Examples
E5.5.1.1	Muconate cycloisomerase
E5.5.1.2	3-carboxy-cis,cis-muconate cycloisomerase
E5.5.1.3	Tetrahydroxypteridine cycloisomerase
E5.5.1.4	Inositol-3-phosphate synthase
E5.5.1.5	Carboxy-cis,cis-muconate cyclase
E5.5.1.6	Chalcone isomerase
E5.5.1.7	Chloromuconate cycloisomerase
E5.5.1.8	(+)-bornyl diphosphate synthase
E5.5.1.9	Cycloeucalenol cycloisomerase
E5.5.1.10	Alpha-pinene-oxide decyclase
E5.5.1.11	Dichloromuconate cycloisomerase
E5.5.1.12	Copalyl diphosphate synthase
E5.5.1.13	Ent-copalyl diphosphate synthase
E5.5.1.14	Syn-copalyl-diphosphate synthase
E5.5.1.15	Terpentedienyl-diphosphate synthase
E5.5.1.16	Halimadienyl-diphosphate synthase
E5.5.1.17	(S)-beta-macrocarpene synthase
E5.5.1.18	Lycopene epsilon-cyclase
E5.5.1.19	Lycopene beta-cyclase
E5.5.1.20	Prosolanapyrone-III cycloisomerase
E5.5.1.n1	D-ribose pyranase

Mechanisms of Isomerases

Ring Expansion and Contraction via Tautomers

A classiexample of ring opening and contraction is the isomerization of glucose (an aldehyde with a six-membered ring) to fructose (a ketone with a five-membered ring). The conversion of D-glucose-6-phosphate to D-fructose-6-phosphate is catalyzed by glucose-6-phosphate isomerase, an intramolecular oxidoreductase. The overall reaction involves the opening of the ring to form an aldose via acid/base catalysis and the subsequent formation of a cis-endiol intermediate. A ketose is then formed and the ring is closed again.

The isomerization of glucose-6-phosphate by glucose-6-phosphate isomerase

Glucose-6-phosphate first binds to the active site of the isomerase. The isomerase opens the ring: its His388 residue protonates the oxygen on the glucose ring (and thereby breaking the O5-C1 bond) in conjunction with Lys518 deprotonating the C1 hydroxyl oxygen. The ring opens to form a straight-chain aldose with an acidiC2 proton. The C3-C4 bond rotates and Glu357 (assisted by His388) depronates C2 to form a double bond between C1 and C2. A cis-endiol intermediate is created and the C1 oxygen is protonated by the catalytiresidue, accompanied by the deprotonation of the endiol C2 oxygen. The straight-chain ketose is formed. To close the fructose ring, the reverse of ring opening occurs and the ketose is protonated.

Epimerization

The conversion of ribulose-5-phosphate to xylulose-5-phosphate

An example of epimerization is found in the Calvin cycle when D-ribulose-5-phosphate is converted into D-xylulose-5-phosphate by ribulose-phosphate 3-epimerase. The substrate and product differ only in stereochemistry at the third carbon in the chain. The underlying mechanism involves the deprotonation of that third carbon to form a reactive enolate intermediate. The enzyme's active site contains two Asp residues. After the substrate binds to the enzyme, the first Asp deprotonates the third carbon from one side of the molecule. This leaves a planar sp²-hybridized intermediate. The second Asp is located on the opposite side of the active side and it protonates the molecule, effectively adding a proton from the back side. These coupled steps invert stereochemistry at the third carbon.

Intramolecular Transfer

A proposed mechanism for chorismate mutase. Clark, T., Stewart, J.D. and Ganem, B. Transition-state analogue inhibitors of chlorismate mutase. Tetrahedron 46 (1990) 731-748. © IUBMB 2001

Chorismate mutase is an intramolecular transferase and it catalyzes the conversion of chorismate to prephenate, used as a precursor for L-tyrosine and L-phenylalanine in some plants and bacteria. This reaction is a Claisen rearrangement that can proceed with or without the isomerase, though the rate increases 10^6 fold in the presence of chorismate mutase. The reaction goes through a chair transition state with the substrate in a trans-diaxial position. Experimental evidence indicates that the isomerase selectively binds the chair transition state, though the exact mechanism of catalysis is not known. It is thought that this binding stabilizes the transition state through electrostatieffects, accounting for the dramatiincrease in the reaction rate in the presence of the mutase or upon addition of a specifically-placed cation in the active site.

Intramolecular Oxidoreduction

Conversion by IPP isomerase

Isopentenyl-diphosphate delta isomerase type I (also known as IPP isomerase) is seen in cholesterol synthesis and in particular it catalyzes the conversion of isopentenyl diphosphate (IPP) to dimethylallyl diphosphate (DMAPP). In this isomerization reaction a stable carbon-carbon double bond is rearranged top create a highly electrophiliallyliisomer. IPP isomerase catalyzes this reaction by the stereoselective antarafacial transposition of a single proton. The double bond is protonated at C4 to form a tertiary carbocation intermediate at C3. The adjacent carbon, C2, is deprotonated from the opposite face to yield a double bond. In effect, the double bond is shifted over.

The Role of Isomerase in Human Disease

Isomerase plays a role in human disease. Deficiencies of this enzyme can cause disorders in humans.

Phosphohexose Isomerase Deficiency

Phosphohexose Isomerase Dificiency (PHI) is also known as phosphoglucose isomerase deficiency or Glucose-6-phosphate isomerase deficiency, and is a hereditary enzyme deficiency. PHI is the second most frequent erthoenzyopathy in glycolysis besides pyruvate kinase deficiency, and is associated with non-spherocytihaemolytianaemia of variable severity. This disease is centered on the glucose-6-phosphate protein. This protein can be found in the secretion of some cancer cells. PHI is the result of a dimerienzyme that catalyses the reversible interconversion of fructose-6-phosphate and gluose-6-phosphate.

PHI is a very rare disease with only 50 cases reported in literature to date.

Diagnosis is made on the basis of the clinical picture in association with biochemical studies revealing erythrocyte GPI deficiency (between 7 and 60% of normal) and identification of a mutation in the GPI gene by molecular analysis.

The deficiency of phosphohexose isomerase can lead to a condition referred to as hemolytisyndrome. As in humans, the hemolytisyndrome, which is characterized by a diminished erythrocyte number, lower hematocrit, lower hemoglobin, higher number of reticulocytes and plasma bilirubin concentration, as well as increased liver- and spleen-somatiindices, was exclusively manifested in homozygous mutants.

Triosephosphate Isomerase Deficiency

The disease referred to as triosephosphate isomerase deficiency (TPI), is a severe autosomal recessive inherited multisystem disorder of glycolyimetabolism. It is characterized by hemolytianemia and neurodegeneration, and is caused by anaerobimetabolidysfunction. This dysfunction results from a missense mutation that effects the encoded TPI protein. The most common mutation is the substitution of gene, Glu104Asp, which produces the most severe phenotype, and is responsible for approximately 80% of clinical TPI deficiency.

TPI deficiency is very rare with less than 50 cases reported in literature. Being an autosomal recessive inherited disease, TPI deficiency has a 25% recurrence risk in the case of heterozygous parents. It is a congenital disease that most often occurs with hemolytianemia and manifests with jaundice. Most patients with TPI for Glu104Asp mutation or heterozygous for a TPI null allele and Glu104Asp have a life expectancy of infancy to early childhood. TPI patients with other mutations generally show longer life expectancy. To date, there are only two cases of individuals with TPI living beyond the age of 6. These cases involve two brothers from Hungary, one who did not develop neurological symptoms until the age of 12, and the older brother who has no neurological symptoms and suffers from anemia only.

Individuals with TPI show obvious symptoms after 6–24 months of age. These symptoms include: dystonia, tremor, dyskinesia, pyramidal tract signs, cardiomyopathy and spinal motor neuron involvement. Patients also show frequent respiratory system bacterial infections.

TPI is detected through deficiency of enzymatiactivity and the build-up of dihyroxyacetone phosphate(DHAP), which is a toxisubstrate, in erythrocytes. This can be detected through physical examination and a series of lab work. In detection, there is generally myopathichanges seen in muscles and chroniaxonal neuropathy found in the nerves. Diagnosis of TPI can be confirmed through molecular genetics. Chorionivillus DNA analysis or analysis of fetal red cells can be used to detect TPI in antenatal diagnosis.

Treatment for TPI is not specific, but varies according to different cases. Because of the range of symptoms TPI causes, a team of specialist may be needed to provide treatment to a single individual. That team of specialists would consists of pediatricians, cardiologists, neurologists, and other healthcare professionals, that can develop a comprehensive plan of action.

Supportive measures such as red cell transfusions in cases of severe anaemia can be taken to treat TPI as well. In some cases, spleen removal (splenectomy) may improve the anaemia. There is no treatment to prevent progressive neurological impairment of any other non-haematological clinical manifestation of the diseases.

Industrial Applications

By far the most common use of isomerases in industrial applications is in sugar manufacturing. Glucose isomerase (also known as xylose isomerase) catalyzes the conversion of D-xylose and D-glucose to D-xylulose and D-fructose. Like most sugar isomerases, glucose isomerase catalyzes the interconversion of aldoses and ketoses.

The conversion of glucose to fructose is a key component of high-fructose corn syrup production. Isomerization is more specifithan older chemical methods of fructose production, resulting in a higher yield of fructose and no side products. The fructose produced from this isomerization reaction is purer with no residual flavors from contaminants. High-fructose corn syrup is preferred by many confectionery and soda manufacturers because of the high sweetening power of fructose (twice that of sucrose), its relatively low cost and its inability to crystallize. Fructose is also used as a sweetener for use by diabetics. Major issues of the use of glucose isomerase involve its inactivation at higher temperatures and the requirement for a high pH (between 7.0 and 9.0) in the reaction environment. Moderately high temperatures, above 70 °C, increase the yield of fructose by at least half in the isomerization step. The enzyme requires a divalent cation such as Co^{2+} and Mg^{2+} for peak activity, an additional cost to manufacturers. Glucose isomerase also has a much higher affinity for xylose than for glucose, necessitating a carefully controlled environment.

The isomerization of xylose to xylulose has its own commercial applications as interest in biofuels has increased. This reaction is often seen naturally in bacteria that feed on decaying plant matter. Its most common industrial use is in the production of ethanol, achieved by the fermentation of xylulose. The use of hemicellulose as source material is very common. Hemicellulose contains xylan, which itself is composed of xylose in $\beta(1,4)$ linkages. The use of glucose isomerase very efficiently converts xylose to xylulose, which can then be acted upon by fermenting yeast. Overall, extensive research in genetiengineering has been invested into optimizing glucose isomerase and facilitating its recovery from industrial applications for re-use.

Glucose isomerase is able to catalyze the isomerization of a range of other sugars, including D-ribose, D-allose and L-arabinose. The most efficient substrates are those similar to glucose and xylose, having equatorial hydroxyl groups at the third and fourth carbons. The current model for the mechanism of glucose isomerase is that of a hydride shift based on X-ray crystallography and isotope exchange studies.

DNA Glycosylase

DNA glycosylases are a family of enzymes involved in base excision repair, classified under Enumber E3.2.2. Base excision repair is the mechanism by which damaged bases in DNA are removed and replaced. DNA glycosylases catalyze the first step of this process. They remove the damaged nitrogenous base while leaving the sugar-phosphate backbone intact, creating an apurinic/apyrimidinisite, commonly referred to as an AP site. This is accomplished by flipping the damaged base out of the double helix followed by cleavage of the N-glycosidibond.

Glycosylases were first discovered in bacteria, and have since been found in all kingdoms of life. In addition to their role in base excision repair DNA glycosylase enzymes have been implicated in the repression of gene silencing in *A. thaliana*, *N. tabacum* and other plants by active demethylation. 5-methylcytosine residues are excised and replaced with unmethylated cytosines allowing access to the chromatin structure of the enzymes and proteins necessary for transcription and subsequent translation.

Monofunctional Vs. Bifunctional Glycosylases

There are two main classes of glycosylases: monofunctional and bifunctional. Monofunctional glycosylases have only glycosylase activity, whereas bifunctional glycosylases also possess AP lyase activity that permits them to cut the phosphodiester bond of DNA, creating a single-strand break without the need for an AP endonuclease. β-Elimination of an AP site by a glycosylase-lyase yields a 3' α,β-unsaturated aldehyde adjacent to a 5' phosphate, which differs from the AP endonuclease cleavage product. Some glycosylase-lyases can further perform δ-elimination, which converts the 3' aldehyde to a 3' phosphate.

Biochemical Mechanism

The first crystal structure of a DNA glycosylase was obtained for E. coli Nth. This structure revealed that the enzyme flips the damaged base out of the double helix into an active site pocket in order to excise it. Other glycosylases have since been found to follow the same general paradigm, including human UNG pictured below. To cleave the N-glycosidibond, monofunctional glycosylases use an activated water molecule to attack carbon 1 of the substrate. Bifunctional glycosylases, instead, use an amine residue as a nucleophile to attack the same carbon, going through a Schiff base intermediate.

Types of Glycosylases

Crystal structures of many glycosylases have been solved. Based on structural similarity, glycosylases are grouped into four superfamilies. The UDG and AAG families contain small, compact glycosylases, whereas the MutM/Fpg and HhH-GPD families comprise larger enzymes with multiple domains.

A wide variety of glycosylases have evolved to recognize different damaged bases. The table below summarizes the properties of known glycosylases in commonly studied model organisms.

Glycosylases in bacteria, yeast and humans					
E. coli	B. cereus	Yeast (S. cerevisiae)	Human	Type	Substrates
AlkA	AlkE	Mag1	MPG	monofunctional	3-meA, hypoxanthine
UDG		Ung1	UNG	monofunctional	uracil
Fpg		Ogg1	hOGG1	bifunctional	8-oxoG, FapyG
Nth		Ntg1	hNTH1	bifunctional	Tg, hoU, hoC, urea, FapyG
	Ntg2				
Nei		Not present	hNEIL1	bifunctional	Tg, hoU, hoC, urea, FapyG, FapyA
	hNEIL2		AP site, hoU		
	hNEIL3		unknown		
MutY		Not present	hMYH	monofunctional	A:8-oxoG
Not present		Not present	hSMUG1	monofunctional	U, hoU, hmU, fU
Not present		Not present	TDG	monofunctional	T:G mispair
Not present		Not present	MBD4	monofunctional	T:G mispair
AlkC	AlkC	Not present	Not present	monofunctional	Alkylpurine
AlkD	AlkD	Not present	Not present	monofunctional	Alkylpurine

DNA glycosylases can be grouped into the following categories based on their substrate(s):

Uracil DNA Glycosylases

In molecular biology, the protein family, Uracil-DNA glycosylase (UDG) is an enzyme that reverts mutations in DNA. The most common mutation is the deamination of cytosine to uracil. UDG repairs these mutations. UDG is crucial in DNA repair, without it these mutations may lead to cancer.

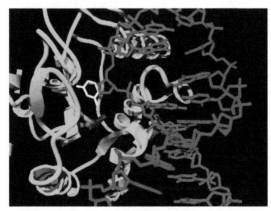

Structure of the base-excision repair enzyme uracil-DNA glycosylase. The uracil residue is shown in yellow.

This entry represents various uracil-DNA glycosylases and related DNA glycosylases (EC), such as uracil-DNA glycosylase, thermophiliuracil-DNA glycosylase, G:T/U mismatch-specifiDNA glycosylase (Mug), and single-strand selective monofunctional uracil-DNA glycosylase (SMUG1).

Uracil DNA glycosylases remove uracil from DNA, which can arise either by spontaneous deamination of cytosine or by the misincorporation of dU opposite dA during DNA replication. The prototypical member of this family is E. coli UDG, which was among the first glycosylases discovered. Four different uracil-DNA glycosylase activities have been identified in mammalian cells, including UNG, SMUG1, TDG, and MBD4. They vary in substrate specificity and subcellular localization. SMUG1 prefers single-stranded DNA as substrate, but also removes U from double-stranded DNA. In addition to unmodified uracil, SMUG1 can excise 5-hydroxyuracil, 5-hydroxymethyluracil and 5-formyluracil bearing an oxidized group at ring C5. TDG and MBD4 are strictly specififor double-stranded DNA. TDG can remove thymine glycol when present opposite guanine, as well as derivatives of U with modifications at carbon 5. Current evidence suggests that, in human cells, TDG and SMUG1 are the major enzymes responsible for the repair of the U:G mispairs caused by spontaneous cytosine deamination, whereas uracil arising in DNA through dU misincorporation is mainly dealt with by UNG. MBD4 is thought to correct T:G mismatches that arise from deamination of 5-methylcytosine to thymine in CpG sites. MBD4 mutant mice develop normally and do not show increased cancer susceptibility or reduced survival. But they acquire more T mutations at CpG sequences in epithelial cells of the small intestine.

The structure of human UNG in complex with DNA revealed that, like other glycosylases, it flips the target nucleotide out of the double helix and into the active site pocket. UDG undergoes a conformational change from an "open" unbound state to a "closed" DNA-bound state.

History

Lindahl was the first to observe repair of uracil in DNA. UDG was purified from *Escherichia coli*, and this hydrolysed the N-glycosidibond connecting the base to the deoxyribose sugar of the DNA backbone.

Function

The function of UDG is to remove mutations in DNA, more specifically removing uracil.

Structure

These proteins have a 3-layer alpha/beta/alpha structure. The polypeptide topology of UDG is that of a classialpha/beta protein. The structure consists primarily of a central, four-stranded, all parallel beta sheet surrounded on either side by a total of eight alpha helices and is termed a parallel doubly wound beta sheet.

Mechanism

Uracil-DNA glycosylases are DNA repair enzymes that excise uracil residues from DNA by cleaving the N-glycosydibond, initiating the base excision repair pathway. Uracil in DNA can arise either through the deamination of cytosine to form mutageniU:G mispairs, or through the incorporation of dUMP by DNA polymerase to form U:A pairs. These aberrant uracil residues are genotoxic.

Localisation

In eukaryoticells, UNG activity is found in both the nucleus and the mitochondria. Human UNG1 protein is transported to both the mitochondria and the nucleus.

Conservation

The sequence of uracil-DNA glycosylase is extremely well conserved in bacteria and eukaryotes as well as in herpes viruses. More distantly related uracil-DNA glycosylases are also found in poxviruses. The N-terminal 77 amino acids of UNG1 seem to be required for mitochondrial localization, but the presence of a mitochondrial transit peptide has not been directly demonstrated. The most N-terminal conserved region contains an aspartiacid residue which has been proposed, based on X-ray structures to act as a general base in the catalytimechanism.

Family

There are two UDG families, named Family 1 and Family 2. Family 1 is active against uracil in ssDNA and dsDNA. Family 2 excise uracil from mismatches with guanine.

Glycosylases of Oxidized Bases

A variety of glycosylases have evolved to recognize oxidized bases, which are commonly formed by reactive oxygen species generated during cellular metabolism. The most abundant lesions formed at guanine residues are 2,6-diamino-4-hydroxy-5-formamidopyrimidine (FapyG) and 8-oxoguanine. Due to mispairing with adenine during replication, 8-oxoG is highly mutagenic, resulting in G to T transversions. Repair of this lesion is initiated by the bifunctional DNA glycosylase OGG1, which recognizes 8-oxoG paired with C. hOGG1 is a bifunctional glycosylase that belongs to the helix-hairpin-helix (HhH) family. MYH recognizes adenine mispaired with 8-oxoG but excises the A, leaving the 8-oxoG intact. OGG1 knockout mice do not show an increased tumor incidence, but accumulate 8-oxoG in the liver as they age. A similar phenotype is observed with the inactivation of MYH, but simultaneous inactivation of both MYH and OGG1 causes 8-oxoG accumulation in multiple tissues including lung and small intestine. In humans, mutations in MYH are associated with increased risk of developing colon polyps and colon cancer. In addition to OGG1 and MYH, human cells contain three additional DNA glycosylases, NEIL1, NEIL2, and NEIL3. These are homologous to bacterial Nei, and their presence likely explains the mild phenotypes of the OGG1 and MYH knockout mice.

8-oxoG (*syn*) in a Hoogsteen base pair with dA (*anti*)

Glycosylases of Alkylated Bases

This group includes E. coli AlkA and related proteins in higher eukaryotes. These glycosylases are monofunctional and recognize methylated bases, such as 3-methyladenine.

AlkA

AlkA refers to 3-methyladenine DNA glycosylase II.

Pathology

- DNA glycosylases involved in base excision repair (BER) may be associated with cancer risk in BRCA1 and BRCA2 mutation carriers.

Epigenetideficiencies in Cancers

Epigenetialterations (epimutations) in DNA glycosylase genes have only recently begun to be evaluated in a few cancers, compared to the numerous previous studies of epimutations in genes acting in other DNA repair pathways (such as MLH1 in mismatch repair and MGMT in direct reversal). Two examples of epimutations in DNA glycosylase genes that occur in cancers are summarized below.

MBD4

MBD4 (methyl-CpG-binding domain protein 4) is a glycosylase employed in an initial step of base excision repair. MBD4 protein binds preferentially to fully methylated CpG sites and to the altered DNA bases at those sites. These altered bases arise from the frequent hydrolysis of cytosine to uracil and hydrolysis of 5-methylcytosine to thymine, producing G:U and G:T base pairs. If the improper uracils or thymines in these base pairs are not removed before DNA replication, they will cause transition mutations. MBD4 specifically catalyzes the removal of T and U paired with guanine (G) within CpG sites. This is an important repair function since about 1/3 of all intragenisingle base pair mutations in human cancers occur in CpG dinucleotides and are the result of G:to A:T transitions. These transitions comprise the most frequent mutations in human cancer. For example, nearly 50% of somatimutations of the tumor suppressor gene p53 in colorectal cancer are G:to A:T transitions within CpG sites. Thus, a decrease in expression of MBD4 could cause an increase in carcinogenimutations.

Hydrolysis of cytosine to uracil

MBD4 expression is reduced in almost all colorectal neoplasms due to methylation of the promoter region of MBD4. Also MBD4 is deficient due to mutation in about 4% of colorectal cancers,

A majority of histologically normal fields surrounding neoplastigrowths (adenomas and colon cancers) in the colon also show reduced MBD4 mRNA expression (a field defect) compared to histologically normal tissue from individuals who never had a colonineoplasm. This finding suggests that epigenetisilencing of MBD4 is an early step in colorectal carcinogenesis.

In a Chinese population that was evaluated, the MBD4 Glu346Lys polymorphism was associated with about a 50% reduced risk of cervical cancer, suggesting that alterations in MBD4 is important in this cancer.

NEIL1

Nei-like (NEIL) 1 is a DNA glycosylase of the Nei family (which also contains NEIL2 and NEIL3). NEIL1 is a component of the DNA replication complex needed for surveillance of oxidized bases before replication, and appears to act as a "cowcatcher" to slow replication until NEIL1 can act as a glycosylase and remove the oxidatively damaged base.

NEIL1 protein recognizes (targets) and removes certain oxidatively-damaged bases and then incises the abasisite via β,δ elimination, leaving 3' and 5' phosphate ends. NEIL1 recognizes oxidized pyrimidines, formamidopyrimidines, thymine residues oxidized at the methyl group, and both stereoisomers of thymine glycol. The best substrates for human NEIL1 appear to be the hydantoin lesions, guanidinohydantoin, and spiroiminodihydantoin that are further oxidation products of 8-oxoG. NEIL1 is also capable of removing lesions from single-stranded DNA as well as from bubble and forked DNA structures. A deficiency in NEIL1 causes increased mutagenesis at the site of an 8-oxo-Gua:pair, with most mutations being G:to T:A transversions.

A study in 2004 found that 46% of primary gastricancers had reduced expression of NEIL1 mRNA, though the mechanism of reduction was not known. This study also found that 4% of gastricancers had mutations in the NEIL1 gene. The authors suggested that low NEIL1 activity arising from reduced expression and/or mutation of the NEIL1 gene was often involved in gastricarcinogenesis.

A screen of 145 DNA repair genes for aberrant promoter methylation was performed on head and neck squamous cell carcinoma (HNSCC) tissues from 20 patients and from head and neck mucosa samples from 5 non-cancer patients. This screen showed that the NEIL1 gene had substantially increased hypermethylation, and of the 145 DNA repair genes evaluated, NEIL1 had the most significantly different frequency of methylation. Furthermore, the hypermethylation corresponded to a decrease in NEIL1 mRNA expression. Further work with 135 tumor and 38 normal tissues also showed that 71% of HNSCtissue samples had elevated NEIL1 promoter methylation.

When 8 DNA repair genes were evaluated in non-small cell lung cancer (NSCLC) tumors, 42% were hypermethylated in the NEIL1 promoter region. This was the most frequent DNA repair abnormality found among the 8 DNA repair genes tested. NEIL1 was also one of six DNA repair genes found to be hypermethylated in their promoter regions in colorectal cancer.

Exoglucanase

Exoglucanase is a type of enzyme of interest for its capability of converting cellulose to useful chemicals, particularly cellulosiethanol.

The main technological impediment to widespread utilization of cellulose for fuels is still the lack of low-cost technologies to convert cellulose. One solution is the use of organisms that are capable of performing this conversion. Development of such organisms, such as Saccharomyces cerevisiae which is capable of secreting high levels of cellobiohydrolases, is already underway. Cellobiohydrolases are exoglucanases derived from fungi .

Function

CBH1, for example, is composed of a carbohydrate binding site, a linker region and a catalytidomain. Once the cellulose chain is bound, it is strung through a tunnel-shaped active site where the cellulose is broken down into two-sugar segments called cellobiose. The structure of the enzyme can be seen in the first figure. The second figure shows the activity of the enzyme, and shows both cellulose binding to the enzyme, as well as the product of this step, cellobiose. Research suggests, however, that the activity of CBH1 is very strong inhibited by the product, cellobiose. Determination of an enzyme that is not as strongly inhibited by the product or finding a way to remove cellobiose from the environment of the enzyme are just more examples of the many challenges that face the use of these enzymes for the creation of biofuels.

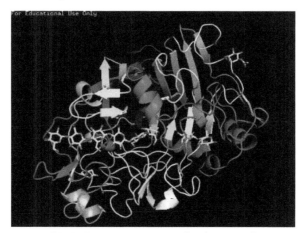

CBH1 Structure, generated using pymol

After this step, the process for creating ethanol is as follows: 3. Separation of sugars from other plant material. 4. Microbial fermentation of the sugar solution to create alcohol. 5. Distillation to purify the products and produce roughly 9% pure alcohol 6. Further purification to bring the ethanol purity to roughly 99.5%

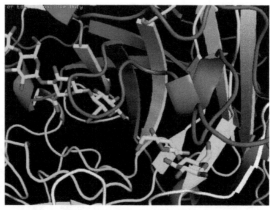

CBH1 zoomed in on the active site where cellulose is cleaved into cellobiose, generated using pymol.

Some notable improvements have been made in this area as well. For example, a strain of yeast capable of producing its own cellulose digesting enzyme has been developed, which would allow the cellulose degradation and the fermentation steps could be at once. This is an important development in the sense that it makes large scale, industrial applications more feasible.

ER Oxidoreductin

ER oxidoreductin 1 (Ero1) is an oxidoreductase enzyme that catalyses the formation and isomerization of protein disulfide bonds in the endoplasmireticulum (ER) of eukaryotes. ER Oxidoreductin 1 (Ero1) is a conserved, luminal, glycoprotein that is tightly associated with the ER membrane, and is essential for the oxidation of protein dithiols. Since disulfide bond formation is an oxidative process, the major pathway of its catalysis has evolved to utilise oxidoreductases, which become reduced during the thiol-disulfide exchange reactions that oxidise the cysteine thiol groups of nascent polypeptides. Ero1 is required for the introduction of oxidising equivalents into the ER and their direct transfer to protein disulfide isomerase (PDI), thereby ensuring the correct folding and assembly of proteins that contain disulfide bonds in their native state.

Homologues of the Saccharomyces cerevisiae Ero1 proteins have been found in all eukaryotiorganisms examined, and contain seven cysteine residues that are absolutely conserved, including three that form the sequence Cys–X–X–Cys–X–X–Cys (where X can be any residue).

The Mechanism of Thiol–Disulfide Exchange between Oxidoreductases

The mechanism of thiol–disulfide exchange between oxidoreductases is understood to begin with the nucleophiliattack on the sulfur atoms of a disulfide bond in the oxidised partner, by a thiolate anion derived from a reactive cysteine in a reduced partner. This generates mixed disulfide intermediates, and is followed by a second, this time intramolecular, nucleophiliattack by the remaining thiolate anion in the formerly reduced partner, to liberate both oxidoreductases. The balance of evidence discussed thus far supports a model in which oxidising equivalents are sequentially transferred from Ero1 via a thiol–disulfide exchange reaction to PDI, with PDI then undergoing a thiol–disulfide exchange with the nascent polypeptide, thereby enabling the formation of disulfide bonds within the nascent polypeptide.

Endoglycosidase

An Endoglycosidase is an enzyme that releases oligosaccharides from glycoproteins or glycolipids. It may also cleave polysaccharide chains between residues that are not the terminal residue, although releasing oligosaccharides from conjugated protein and lipid molecules is more common.

It breaks the glycosidibonds between two sugar monomer in the polymer. It is different from exoglycosidase that it does not do so at the terminal residue. Hence, it is used to release long carbohydrates from conjugated molecules. If an exoglycosidase were used, every monomer in the polymer would have to be removed, one by one from the chain, taking a long time. An endoglycosidase cleaves, giving a polymeriproduct.

$$PROTEIN-x_1-x_2-x_3-x_4-x_5-x_6-x_7-x_8-x_9-x_{10}-x_{11}-...-x_n$$

Mechanism Overview

Examples of various endoglycosidases		
Endoglycosidase	Gylcoside	Bond hydrolysed
D		
F	Glc-Nac	Gl// Nac
F1		
F2	Glc-Nac	Gl// Nac
H	diacetylchitobiose	Na// asparagine
Nac: N-Acetylglucosamine		

The mechanism is an enzymatihydrolysis that requires two critical molecules; a proton donor (most likely an acid) and a nucleophile(most likely a base). The Endoglycosidases mechanism has two forms; an acid catalyzed protonation of the glycosidioxygen yielding stereochemical retention at the anomericarbon or an acid catalyzed protonation of the glycosidioxygen with a concomitant attack of a water molecule activated by the base residue yielding a stereochemical inversion.

Both mechanisms exhibit the same distance between the proton donor and the glycosidioxygen, situating the proton donor close enough to the glycosidioxygen for hydrogen bonding. It is the distance between the nucleophile and the anomericcarbon where the two mechanisms begin to diverge. Because the inversion mechanism must accommodate enough space for the water molecule, the nucleophile is situated further away from the anomericcarbon. In the retention mechanism, this distance is only 5.5 -7 angstroms but increases to 9-10 angstroms in the inversion mechanism. Furthermore, the inversion mechanism was found to proceed through a single displacement mechanism involving an oxocarbenium ion-like transition state. Due to the retention mechanism's proximity between the two carboxyl groups, it goes through a double displacement mechanism that produces a covalent glycosyl-enzyme intermediate.

A exoglycosidase would remove each carbohydrate monomer (x) one by one from the end, starting at x_n, whereas and endogylcosidase can cut at any glycosidibond (-) and may cleave after a signature 'link oligosaccharide' that links certain carbohydrates to certain proteins.

Applications and Potential Uses

There has been great potential shown in the use of endoglycosidase enzymes undergoing mutagenesis. This new mutated enzyme when exposed to the proper compounds will undergo oligosaccharide synthesis and will not hydrolyze the newly formed polymer chains. This is an extremely useful tool, as oligosaccharides have a great potential for use as therapeutics. For example, globo H hexasaccharide will indicate cancer related malignant cell transformation in the breast, prostate and ovaries.

Endoglycosidases also have potential application in fighting autoimmune diseases such as arthritis and systemilupus erythematosus. In 2008, a team of researchers demonstrated that injection of endgoglycosidase S "efficiently removes the IgG-associated sugar domain in vivo and interferes with autoantibody-mediated proinflammatory processes in a variety of autoimmune models." Clearly the manipulation and mutation of this enzyme holds great promise for being able to fight a variety of diseases in the body.

Endoenzyme

An endoenzyme, or intracellular enzyme, is an enzyme that functions within the cell in which it was produced. Because the majority of enzymes fall within this category, the term is used primarily to differentiate a specifienzyme from an exoenzyme. It is possible for a single enzyme to have both endoenzymatiand exoenzymatifunctions.

Example:Glycolytienzymes,enzymes of Kreb's Cycle. enzymes are a type of protein that speed up chemical reactions in cells. enzymes are specifito the job they do. only molecules that are the correct shape can fit into the enzyme. this is called the lock and key model. enzymes work outside of the cell (extracellular enzymes) as well as inside the cell (intracellular enzymes) In most cases the term endoenzyme refers to an enzyme that binds to a bond 'within the body' of a large molecule - usually a polymer. For example an endoamylase would break down large amylose molecules into shorter dextrin chains. On the other hand, an exoenzyme removes subunits from the polymer one at a time from one end; in effect it can only act at the end ponts of a polymer. An exoamylase would therefore remove one glucose molecule at a time from the end of an amylose molecule.

Exoenzyme

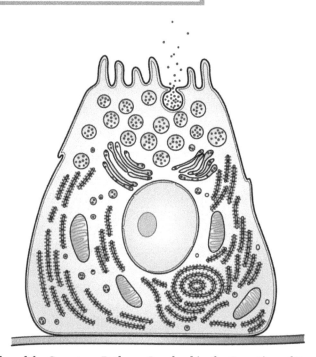

Organelles of the Secretory Pathway Involved in the Secretion of Exoenzymes

An exoenzyme, or extracellular enzyme, is an enzyme that is secreted by a cell and functions outside of that cell. Exoenzymes are produced by both prokaryotiand eukaryoticells and have been shown to be a crucial component of many biological processes. Most often these enzymes are involved in the breakdown of larger macromolecules. The breakdown of these larger macromolecules is critical for allowing their constituents to pass through the cell membrane and enter into the cell. For humans and other complex organisms, this process is best characterized by the digestive system which breaks down solid food via exoenzymes. The small molecules, generated by the exoenzyme activity, enter into cells and are utilized for various cellular functions. Bacteria and fungi also produce exoenzymes to digest nutrients in their environment, and these organisms can be used to conduct laboratory assays to identify the presence and function of such exoenzymes. Some pathogenispecies also use exoenzymes as virulence factors to assist in the spread of these disease causing microorganisms. In addition to the integral roles in biological systems, different classes of microbial exoenzymes have been used by humans since pre-historitimes for such diverse purposes as food production, biofuels, textile production and in the paper industry. Another important role that microbial exoenzymes serve is in the natural ecology and bioremediation of terrestrial and marine environments.

History

Very limited information is available about the original discovery of exoenzymes. According to Mcrriam-Webster dictionary, the term "exoenzyme" was first recognized in the English language in 1908. The book "Intracellular Enzymes: A Course of Lectures Given in the Physiological," by Horace Vernon is thought to be the first publication using this word in that year. Based on the book, it can be assumed that the first known exoenzymes were pepsin and trypsin, as both are mentioned by Vernon to have been discovered by scientists Briike and Kiihne before 1908.

Function

In bacteria and fungi, exoenzymes play an integral role in allowing the organisms to effectively interact with their environment. Many bacteria use digestive enzymes to break down nutrients in their surroundings. Once digested, these nutrients enter the bacterium, where they are used to power cellular pathways with help from endoenzymes.

Many exoenzymes are also used as virulence factors. Pathogens, both bacterial and fungal, can use exoenzymes as a primary mechanism with which to cause disease. The metaboliactivity of the exoenzymes allows the bacterium to invade host organisms by breaking down the host cells' defensive outer layers or by necrotizing body tissues of larger organisms. Many gram-negative bacteria have injectisomes, or flagella-like projections, to directly deliver the virulent exoenzyme into the host cell using a type three secretion system. With either process, pathogens can attack the host cell's structure and function, as well as its nucleiDNA.

In eukaryoticells, exoenzymes are manufactured like any other enzyme via protein synthesis, and are transported via the secretory pathway. After moving through the rough endoplasmireticulum, they are processed through the Golgi apparatus, where they are packaged in vesicles and released out of the cell. In humans, a majority of such exoenzymes can be found in the digestive system and are used for metabolibreakdown of macronutrients via hydrolysis. Breakdown of these nutrients allows for their incorporation into other metabolipathways.

Examples of Exoenzymes as Virulence Factors

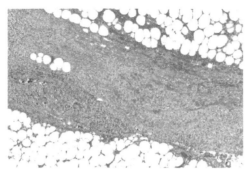

Microscopiview of necrotizing fasciitis as caused by Streptococcus pyogenes

Necrotizing Enzymes

Necrotizing enzymes destroy cells and tissue. One of the best known examples is an exoenzyme produced by Streptococcus pyogenes that causes necrotizing fasciitis in humans.

Coagulase

By binding to prothrombin, coagulase facilitates clotting in a cell by ultimately converting fibrinogen to fibrin. Bacteria such as Staphylococcus aureus use the enzyme to form a layer of fibrin around their cell to protect against host defense mechanisms.

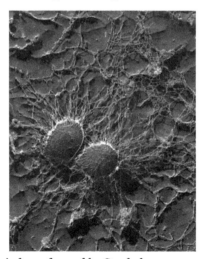

Fibrin layer formed by Staphyloccocus aureus

Kinases

The opposite of coagulase, kinases can dissolve clots. S. aureus can also produce staphylokinase, allowing them to dissolve the clots they form, to rapidly diffuse into the host at the correct time.

Hyaluronidase

Similar to collagenase, hyaluronidase enables a pathogen to penetrate deep into tissues. Bacteria such as Clostridium do so by using the enzyme to dissolve collagen and hyaluroniacid, the protein and saccharides, respectively, that hold tissues together.

Hemolysins

Hemolysins target erythrocytes, or red blood cells. Attacking and lysing these cells allows the pathogen to harm the host organism, and also provides it with a source of iron from the lysed hemoglobin, like the fungus Candida albicans. Organisms can either by alpha-hemolytic, beta-hemolytic, or gamma-hemolyti(non-hemolytic).

Examples of Digestive Exoenzymes

Amylases

Amylases are a group of extracellular enzymes (glycoside hydrolases) that catalyze the hydrolysis of starch. These enzymes are grouped into three classes based on their amino acid sequences, mechanism of reaction, method of catalysis and their structure. The different classes of amylases are α-amylases, β-amylases, and glucoamylases. The α-amylases hydrolyze starch by randomly cleaving the 1,4-a-D-glucosidilinkages between glucose units, β-amylases cleave non-reducing chain ends of components of starch such as amylose, and glucoamylases hydrolyze glucose molecules from the ends of amylose and amylopectin. Amylases are critically important extracellular enzymes and are found in plants, animals and micro-organisms. In humans, amylases are secreted by both the pancreas and salivary glands with both sources of the enzyme required for complete starch hydrolysis.

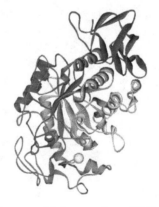

Pancreatialpha-amylase 1HNY

Lipoprotein Lipase

Lipoprotein lipase (LPL) is a type of digestive enzyme that helps regulate the uptake of triacylglycerols from chylomicrons and other low-density lipoproteins from fatty tissues in the body. The exoenzymatifunction allows it to break down the triacylglycerol into two free fatty acids and one molecule of monoacylglycerol. LPL can be found in endothelial cells in fatty tissues, such as adipose, cardiac, and muscle. Lipoprotein lipase is downregulated by high levels of insulin, and upregulated by high levels of glucagon and adrenaline.

Pectinase

Pectinases, also called pectolytienzymes, are a class of exoenzymes that are involved in the breakdown of pectisubstances, most notably pectin. Pectinases can be classified into two different groups based on their action against the galacturonan backbone of pectin: de-esterifying and depolymerizing. These exoenzymes can be found in both plants and microbial organisms including fungi and bacteria. Pectinases are most often used to break down the pectielements found in plants and plant-derived products.

Pepsin

Discovered in 1836, pepsin was one of the first enzymes to be classified as an exoenzyme. The enzyme is first made in the inactive form, pepsinogen by chief cells in the lining of the stomach. With an impulse from the vagus nerve, pepsinogen is secreted into the stomach, where it mixes with hydrochloriacid to form pepsin. Once active, pepsin works to break down proteins in foods such as dairy, meat, and eggs. Pepsin works best at the pH of gastriacid, 1.5 to 2.5, and is deactivated when the acid is neutralized to a pH of 7.

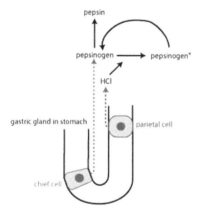

Pepsin is activated in the stomach via secretion of pepsinogen and hydrochloriacid.

Trypsin

Also one of the first exoenzymes to be discovered, trypsin was named in 1876, forty years after pepsin. This enzyme is responsible for the breakdown of large globular proteins and its activity is specifito cleaving the C-terminal sides of arginine and lysine amino acid residues. It is the derivative of trypsinogen, an inactive precursor that is produced in the pancreas. When secreted into the small intestine, it mixes with enterokinase to form active trypsin. Due to its role in the small intestine, trypsin works at an optimal pH of 8.0.

Bacterial Assays

The production of a particular digestive exoenzyme by a bacterial cell can be assessed using plate assays. Bacteria are streaked across the agar, and are left to incubate. The release of the enzyme into the surroundings of the cell cause the breakdown of the macromolecule on the plate. If a reaction does not occur, this means that the bacteria does not create an exoenzyme capable of interacting with the surroundings. If a reaction does occur, it becomes clear that the bacteria does possess an exoenzyme, and which macromolecule is hydrolyzed determines its identity.

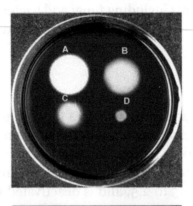

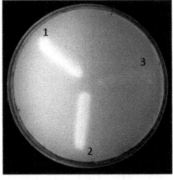

Results of bacterial assays. Left:amylase bacterial assay on a starch medium. A indicates a positive result, D indicates a negative result. Right: lipase bacterial assay on an olive oil medium. 1 shows a positive result, 3 shows a negative result

Amylase

Amylase breaks down carbohydrates into mono- and disaccharides, so a starch agar must be used for this assay. Once the bacteria is streaked on the agar, the plate is flooded with iodine. Since iodine binds to starch but not its digested byproducts, a clear area will appear where the amylase reaction has occurred. Bacillus subtilis is a bacteria that results in a positive assay as shown in the picture.

Lipase

Lipase assays are done using a lipid agar with a spirit blue dye. If the bacteria has lipase, a clear streak will form in the agar, and the dye will fill the gap, creating a dark blue halo around the cleared area. Staphylococcus epidermis results in a positive lipase assay.

Biotechnological and Industrial Applications

Microbiological sources of exoenzymes including amylases, proteases, pectinases, lipases, xylanases, cellulases among others are used for a wide range of biotechnological and industrial uses including biofuel generation, food production, paper manufacturing, detergents and textile production. Optimizing the production of biofuels has been a focus of researchers in recent years and is centered around the use of microorganisms to convert biomass into ethanol. The enzymes that are of particular interest in ethanol production are cellobiohydrolase which solubilizes crystalline cellulose and xylanase that hydrolyzes xylan into xylose. One model of biofuel production is the use of a mixed population of bacterial strains or a consortium that work to facilitate the breakdown of cellulose materials into ethanol by secreting exoenzymes such as cellulases and laccases. In addition to the important role it plays in biofuel production, xylanase is utilized in a number of other industrial and biotechnology applications due to its ability to hydrolyze cellulose and hemicellulose. These applications include the breakdown of agricultural and forestry wastes, working as a feed additive to facilitate greater nutrient uptake by livestock, and as an ingredient in bread making to improve the rise and texture of the bread.

Triglyceride   Methanol (3)   Glycerol   Methyl Esters (3)

GeneriBiodiesel Reaction. Lipases can serve as a biocatalyst in this reaction

Lipases are one of the most used exoenzymes in biotechnology and industrial applications. Lipases make ideal enzymes for these applications because they are highly selective in their activity, they are readily produced and secreted by bacteria and fungi, their crystal structure is well characterized, they do not require cofactors for their enzymatiactivity, and they do not catalyze side reactions. The range of uses of lipases encompasses production of biopolymers, generation of cosmetics, use as a herbicide, and as an effective solvent. However, perhaps the most well known use of lipases in this field is its use in the production of biodiesel fuel. In this role, lipases are used to convert vegetable oil to methyl- and other short-chain alcohol esters by a single transesterification reaction.

Cellulases, hemicellulases and pectinases are different exoenzymes that are involved in a wide variety of biotechnological and industrial applications. In the food industry these exoenzymes are used in the production of fruit juices, fruit nectars, fruit purees and in the extraction of olive oil among many others. The role these enzymes play in these food applications is to partially breakdown the plant cell walls and pectin. In addition to the role they play in food production, cellulases are used in the textile industry to remove excess dye from denim, soften cotton fabrics and restore the color brightness of cotton fabrics. Cellulases and hemicellulases (including xylanases) are also used in the paper and pulp industry to de-ink recycled fibers, modify coarse mechanical pulp and for the partial or complete hydrolysis of pulp fibers. Cellulases and hemicellulases are used in these industrial applications due to their ability to hydrolyze the cellulose and hemicellulose components found in these materials.

Bioremediation Applications

Bioremediation is a process in which pollutants or contaminants in the environment are removed through the use of biological organisms or their products. The removal of these often hazardous pollutants is mostly carried out by naturally occurring or purposely introduced microorganisms that are capable of breaking down or absorbing the desired pollutant. The types of pollutants that are often the targets of bioremediation strategies are petroleum products (including oil and solvents) and pesticides. In addition to the microorganisms ability to digest and absorb the pollutants, their secreted exoenzymes play an important role in many bioremediation strategies.

Water pollution from runoff of soil & fertilizer

Fungi have been shown to be viable organisms to conduct bioremediation and have been used to aid in the decontamination of a number of pollutants including polycycliaromatihydrocarbons (PAHs), pesticides, synthetidyes, chlorophenols, explosives, crude oil and many others. While fungi can breakdown many of these contaminants intracellularly, they also secrete numerous oxidative exoenzymes that work extracellularly. One critical aspect of fungi in regards to bioremediation is that they secrete these oxidative exoenzymes from their ever elongating hyphal tips. Laccases are an important oxidative enzyme that fungi secrete and use oxygen to oxidize many pollutants. Some of the pollutants that laccases have been used to treat include dye-containing effluents from the textile industry, wastewater pollutants (chlorophenols, PAHs, etc.), and sulphur-containing compounds from coal processing.

Exocytivesicles move along actin microfilaments towards the fungal hyphal tip where they release their contents including exoenzymes

Bacteria are also a viable source of exoenzymes capable to facilitating the bioremediation of the environment. There are many examples of the use of bacteria for this purpose and their exoenzymes encompass many different classes of bacterial enzymes. Of particular interest in this field are bacterial hydrolases as they have an intrinsilow substrate specificity and can be used for numerous pollutants including solid wastes. Plastiwastes including polyurethanes are particularly hard to degrade, but an exoenzyme has been identified in a Gram-negative bacterium, *Comamonas acidovorans*, that was capable of degrading polyurethane waste in the environment. Cell-free use of microbial exoenzymes as agents of bioremediation is also possible although their activity is often not as robust and introducing the enzymes into certain environments such as soil has been challenging. In addition to terrestrial based microorganisms, marine based bacteria and their exoenzymes show potential as candidates in the field of bioremediation. Marine based bacteria have been utilized in the removal of heavy metals, petroleum/ diesel degradation and in the removal of polyaromatihydrocarbons among others.

Fucosidase

Tissue alpha-L-fucosidase is an enzyme that in humans is encoded by the *FUCA1* gene.

Alpha-Fucosidase is an enzyme that breaks down fucose.

Fucosidosis is an autosomal recessive lysosomal storage disease caused by defective alpha-L-fucosidase with accumulation of fucose in the tissues. Different phenotypes include clinical features such as neurologideterioration, growth retardation, visceromegaly, and seizures in a severe early form; coarse facial features, angiokeratoma corporis diffusum, spasticity and delayed psychomotor development in a longer surviving form; and an unusual spondylometaphyseoepiphyseal dysplasia in yet another form.[supplied by OMIM]

Glucansucrase

Glucansucrase (also known as glucosyltransferase) is an enzyme in the glycoside hydrolase family GH70 used by lactiacid bacteria to split sucrose and use resulting glucose molecules to build long, sticky biofilm chains. These extracellular homopolysaccharides are called α-glucan polymers.

Glucansucrase enzymes can synthesize a variety of glucans with differing solubilities, rheology, and other properties by altering the type of glycosidilinkage, degree of branching, length, mass, and conformation of the polymers. Glucansucrases are classified according to the glycosidilinkage they catalyze. They can be mutansucrases, dextransucrases, alternansucrases, or reuteransucrases. This versatility has made glucansucrase useful for industrial applications. Glucansucrase's role in cariogenesis is a major point of interest. Glucan polymers stick to teeth in the human mouth and cause tooth decay. There is hope to make dental caries a thing of the past by knocking out this enzyme.

Structure

Glucansucrases are large, extracellular proteins with average molecular masses around 160,000 Daltons. Therefore crystallography studies have only been carried out for fragments of the enzymes, not complete structures. However, glucansucrase is very similar to α-amylase, another sugar-cutting enzyme. Glucansucrase thus has many of the same structural features. For example, both enzymes have three domains in their catalyticore and a $(\beta/\alpha)_8$ barrel.

Glucansucrase has 5 major domains: A, B, C, IV, and V. The domains in glucansucrase, however, have a different arrangement than those in α-amylase. The folding characteristics of α-amylase and glucansucrase are still very similar, but their domains are permuted.

Domains A, B, IV, and V are built from two discontiguous parts of the polypeptide chain, causing the chain to follow a U-shape. From the N- to C-terminus, the polypeptide chain goes in the following order: V, IV, B, A, C, A, B, IV, V. The domain is the only one made up of a continuous polypeptide sequence.

Domain A contains the $(\beta/\alpha)_8$ barrel and the catalytisite. In the catalytisite, three residues in particular play important roles for enzymatiactivity: a nucleophiliaspartate, an acid/base glutamate, and an additional aspartate to stabilize the transition state.

Domain B makes up a twisted antiparallel β sheet. Some of the loops in domain B help shape the groove near the catalytisite. Additionally, some amino acids between domains A and B form a calcium binding site near the nucleophiliaspartate. The Ca^{2+} ion is necessary for enzyme activity.

Reaction and Mechanism

Glucansucrase has two parts to its reaction. First it cleaves a glycosidibond to split sucrose. Products of the reaction are the constituent monosaccharides glucose and fructose. This glucose is added to a growing glucan chain. Glucansucrase uses the energy released from bond cleavage to drive glucan synthesis. Both sucrose breakdown and glucan synthesis occur in the same active site.

The first step is carried out through a transglycosylation mechanism involving a glyco-syl-enzyme intermediate in subsite-1. Glutamate is likely the catalytiacid/base, aspar-tate the nucleophile, and another aspartate the transition state stabilizer. These three residues are all highly conserved and mutating them leads to a significant decrease in enzymatiactivity.

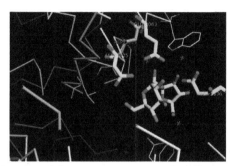

Active site of glucansucrase in Lactobacillus reuteri

The glucansucrase mechanism has historically been controversial in the scientifiliterature. The mechanism involves two displacements. The first originates from a glycosidicleavage of the sucrose substrate between subsites -1 and +1. This releases fructose and forms a sugar-enzyme intermediate when the glucose unit attaches to the nucleophile.

The second displacement is transfer of a glucosyl moiety to an acceptor, such as a grow-ing glucan chain. The debate in the past was over whether the glucosyl group attached to the non-reducing or reducing end of an incoming acceptor. Additional investigations pointed to a non-reducing mechanism with a single active site.

Evolution

Glucansucrase proteins likely evolved from an amylase enzyme precursor. The two enzymes have similar folding patterns and protein domains. In fact, past attempts to produce drugs targeting glucansucrase have not been successful because the drugs also disrupted amylase, which is necessary to break down starches. This occurred because the active sites of the two enzymes are nearly the same. Glucansucrase likely main-tained a highly-conserved active site as it underwent a different evolutionary path.

Health

Glucansucrase allows the oral bacteria *Streptococcus mutans* to metabolize sucrose into lactiacid. This lactiacid lowers the pH around teeth and dissolves calcium phosphate in tooth enamel, leading to tooth decay. Additionally, the synthesis of glucan aids *S. mutans* in adhering to the surface of teeth. As the polymers accumulate, they help more acid-produc-ing bacteria stay on teeth. Consequently, glucansucrase is such an attractive drug target to prevent tooth decay. If *S. mutans* can no longer break down sucrose and synthesize glucan, calcium phosphate is not degraded and bacteria cannot adhere as easily to teeth.

Industry

Bacteria with glucansucrase enzymes are used extensively in industry for a variety of applications. The polymer dextran is one prominent example of a highly used sugar chain. It is fermented at commercial scale for uses in veterinary medicine, separation technology, biotechnology, the food industry for gelling, viscosifying, and emulsifying, in human medicine as a prebiotic, cholesterol-lowering agent or blood plasma expander, and more.

Immobilized Enzyme

An immobilized enzyme is an enzyme that is attached to an inert, insoluble material such as calcium alginate (produced by reacting a mixture of sodium alginate solution and enzyme solution with calcium chloride). This can provide increased resistance to changes in conditions such as pH or temperature. It also allows enzymes to be held in place throughout the reaction, following which they are easily separated from the products and may be used again - a far more efficient process and so is widely used in industry for enzyme catalysed reactions. An alternative to enzyme immobilization is whole cell immobilization.

Enzymes immobilised in beads of alginate gel

Commercial Use

Immobilized enzymes are very important for commercial uses as they possess many benefits to the expenses and processes of the reaction of which include:

- Convenience: Minuscule amounts of protein dissolve in the reaction, so workup can be much easier. Upon completion, reaction mixtures typically contain only solvent and reaction products.

- Economy: The immobilized enzyme is easily removed from the reaction making it easy to recycle the biocatalyst. This is particularly useful in processes such as the production of Lactose Free Milk, as the milk can be drained from a container leaving the enzyme (Lactase) inside ready for the next batch.

- Stability: Immobilized enzymes typically have greater thermal and operational stability than the soluble form of the enzyme.

In the past, biological washing powders and detergents would contain many proteases and lipases which would break down dirt. However, when the cleaning products would come into contact with the skin, it would create allergireactions. This is why immobilization of enzymes are important, not just economically.

Immobilization of An Enzyme

There are various ways by which one can immobilize an enzyme:

- Affinity-tag binding: Enzymes may be immobilized to a surface, e.g. in a porous material, using non-covalent or covalent Protein tags. This technology has been established for protein purification purposes, and has recently been applied for biocatalysis applications by EziG™ with the His-tag. This technique is the only one which can be regarded as generally applicable, and can be performed without prior enzyme purification with a pure preparation as the result. Porous glass and derivatives thereof are used, where the porous surface can be adapted in terms of hydrophobicity to suit the enzyme in question.

- Adsorption on glass, alginate beads or matrix: Enzyme is attached to the outside of an inert material. In general, this method is the slowest among those listed here. As adsorption is not a chemical reaction, the active site of the immobilized enzyme may be blocked by the matrix or bead, greatly reducing the activity of the enzyme.

- Entrapment: The enzyme is trapped in insoluble beads or microspheres, such as calcium alginate beads. However, this insoluble substances hinders the arrival of the substrate, and the exit of products.

- Cross-linkage: Enzyme molecules are covalently bonded to each other to create a matrix consisting of almost only enzyme. The reaction ensures that the binding site does not cover the enzyme's active site, the activity of the enzyme is only affected by immobility. However, the inflexibility of the covalent bonds precludes the self-healing properties exhibited by chemoadsorbed self-assembled monolayers. Use of a spacer molecule like poly(ethylene glycol) helps reduce the sterihindrance by the substrate in this case.

- Covalent bond: The enzyme is bound covalentely to an insoluble support (such as silica gel). This approach provides the strongest enzyme/support interaction, and so the lowest protein leakage during catalysis.

Immobilization of a Substrate for Enzymatic Reactions

Another widely used application of the immobilization approach together with enzymes has been the enzymatireactions on immobilized substrates. This approach facilitates the analysis of enzyme activities and mimics the performance of enzymes on e.g. cell walls.

References

* Boyce S, Tipton KF (2005). "Enzyme Classification and Nomenclature". ELS. doi:10.1038/npg. els.0003893. ISBN 0470016175.

* Hausmann R. To grasp the essence of life: a history of molecular biology. Dordrecht: Springer. pp. 198–199. ISBN 978-90-481-6205-5.

* Lewis R (2008). Human genetics : concepts and applications (8th ed.). Boston: McGraw-Hill/ Higher Education. p. 32. ISBN 978-0-07-299539-8.

* Datta SP, Smith GH, Campbell PN (2000). Oxford Dictionary of Biochemistry and Molecular Biology (Rev. ed.). Oxford: Oxford Univ. Press. ISBN 978-0-19-850673-7.

* Enzyme nomenclature, 1978 recommendations of the Nomenclature Committee of the International Union of Biochemistry on the nomenclature and classification of enzymes. New York: Academic Press. 1979. ISBN 9780323144605.

* McNaught AD (1997). Compendium of Chemical Terminology (2nd ed.). Oxford: Blackwell Scientific Publications. ISBN 0-9678550-9-8.

* Whitesell JK, Fox MA (2004). Organic Chemistry (3rd ed.). Sudbury, Mass.: Jones and Bartlett. pp. 220–222. ISBN 978-0-7637-2197-8.

* Cornish-Bowden A. Fundamentals of Enzyme Kinetics (4th ed.). Weinheim: Wiley-VCH. pp. 238–241. ISBN 978-3-527-66548-8.

* Bugg T (2012). "Chapter 10: Isomerases". Introduction to Enzyme and Coenzyme Chemistry (3rd ed.). Wiley. ISBN 978-1-118-34896-3.

* Lindahl, T. (1986). "DNA Glycosylases in DNA Repair". Mechanisms of DNA Damage and Repair: 335–340. doi:10.1007/978-1-4615-9462-8_36. ISBN 978-1-4615-9464-2.

* "ENZYME Entry: EC 2.7.11.22". ExPASy: Bioinformatics Resource Portal. Swiss Institute of Bioinformatics. Retrieved 4 December 2013.

* "1aqy Summary". Protein Data Bank in Europe Bringing Structure to Biology. The European Bioinformatics Institute. Retrieved 11 December 2013.

* "EC 2.9.1". School of Biological & Chemical Sciences at Queen Mary, University of London. Nomenclature Committee of the International Union of Biochemistry and Molecular Biology (NC-IUBMB). Retrieved 11 December 2013.

* "Succinyl-CoA:3-ketoacid CoA transferase deficiency". Genetics Home Reference. National Institute of Health. Retrieved 4 November 2013.

* "Carnitine plamitoyltransferase I deficiency". Genetics Home Reference. National Institute of Health. Retrieved 4 November 2013.

* "Choline O-Acetyltransferase". GeneCards: The Human Gene Compendium. Weizmann Institute of Science. Retrieved 5 December 2013.

Enzyme: Inhibitors, Activators and Promiscuity

This chapter discusses the topics of enzyme inhibitors, enzyme activators and the phenomena of enzyme promiscuity. Enzyme inhibitors are molecules that adhere to enzymes and limit their activity while enzyme activators are molecules that attach to enzymes and augment their activity. These topics have been discussed in detail and a section of the chapter studies enzyme promiscuity the ability of an enzyme to catalyze a fortuitous side reaction in addition to its main reaction.

Enzyme Inhibitor

An enzyme inhibitor is a molecule that binds to an enzyme and decreases its activity. Since blocking an enzyme's activity can kill a pathogen or correct a metabolic imbalance, many drugs are enzyme inhibitors. They are also used in pesticides. Not all molecules that bind to enzymes are inhibitors; *enzyme activators* bind to enzymes and increase their enzymatic activity, while enzyme substrates bind and are converted to products in the normal catalytic cycle of the enzyme.

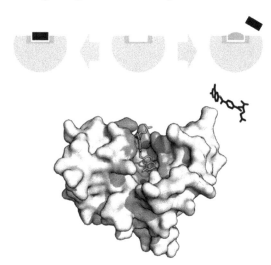

An enzyme binding site that would normally bind substrate can alternatively bind a competitive inhibitor, preventing substrate access. Dihydrofolate reductase is inhibited by methotrexate which prevents binding of its substrate, folic acid. Binding site in blue, inhibitor in green, and substrate in black. (PDB: 4QI9)

The binding of an inhibitor can stop a substrate from entering the enzyme's active site and/or hinder the enzyme from catalyzing its reaction. Inhibitor binding is either reversible or irreversible. Irreversible inhibitors usually react with the enzyme and change it chemically (e.g. via covalent bond formation). These inhibitors modify key amino acid residues needed for enzymatic activity. In contrast, reversible inhibitors bind non-covalently and different types of inhibition are produced depending on whether these inhibitors bind to the enzyme, the enzyme-substrate complex, or both.

Many drug molecules are enzyme inhibitors, so their discovery and improvement is an active area of research in biochemistry and pharmacology. A medicinal enzyme inhibitor is often judged by its specificity (its lack of binding to other proteins) and its potency (its dissociation constant, which indicates the concentration needed to inhibit the enzyme). A high specificity and potency ensure that a drug will have few side effects and thus low toxicity.

Enzyme inhibitors also occur naturally and are involved in the regulation of metabolism. For example, enzymes in a metabolic pathway can be inhibited by downstream products. This type of negative feedback slows the production line when products begin to build up and is an important way to maintain homeostasis in a cell. Other cellular enzyme inhibitors are proteins that specifically bind to and inhibit an enzyme target. This can help control enzymes that may be damaging to a cell, like proteases or nucleases. A well-characterised example of this is the ribonuclease inhibitor, which binds to ribonucleases in one of the tightest known protein–protein interactions. Natural enzyme inhibitors can also be poisons and are used as defences against predators or as ways of killing prey.

Reversible Inhibitors

Types of Reversible Inhibitors

Reversible inhibitors attach to enzymes with non-covalent interactions such as hydrogen bonds, hydrophobic interactions and ionic bonds. Multiple weak bonds between the inhibitor and the active site combine to produce strong and specific binding. In contrast to substrates and irreversible inhibitors, reversible inhibitors generally do not undergo chemical reactions when bound to the enzyme and can be easily removed by dilution or dialysis.

There are four kinds of reversible enzyme inhibitors. They are classified according to the effect of varying the concentration of the enzyme's substrate on the inhibitor.

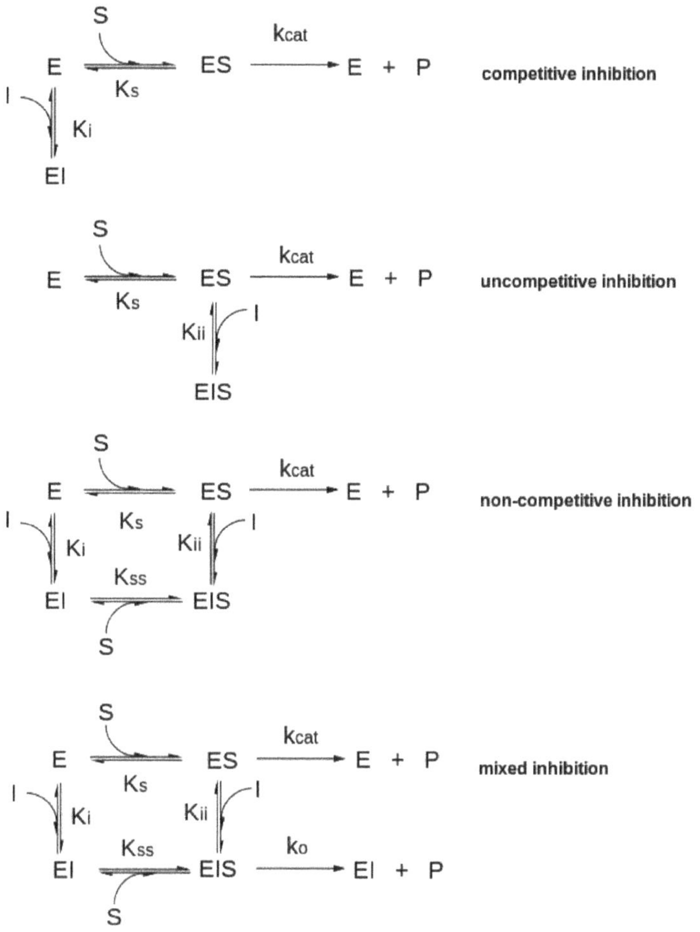

Types of inhibition. This classification was introduced by W.W. Cleland.

- In competitive inhibition, the substrate and inhibitor cannot bind to the enzyme at the same time, as shown in the figure on the right. This usually results from the inhibitor having an affinity for the active site of an enzyme where the substrate also binds; the substrate and inhibitor *compete* for access to the enzyme's active site. This type of inhibition can be overcome by sufficiently high concentrations of substrate (V_{max} remains constant), i.e., by out-competing the inhibitor. However, the apparent K_m will increase as it takes a higher concentration of the substrate to reach the K_m point, or half the V_{max}. Competitive inhib-itors are often similar in structure to the real substrate.

- In uncompetitive inhibition, the inhibitor binds only to the substrate-enzyme complex, it should not be confused with non-competitive inhibitors. This type of inhibition causes V_{max} to decrease (maximum velocity decreases as a result of removing activated complex) and K_m to decrease (due to better binding efficiency as a result of Le Chatelier's principle and the effective elimination of the ES complex thus decreasing the K_m which indicates a higher binding affinity).

- In non-competitive inhibition, the binding of the inhibitor to the enzyme reduces its activity but does not affect the binding of substrate. As a result, the extent of inhibition depends only on the concentration of the inhibitor. V_{max} will decrease due to the inability for the reaction to proceed as efficiently, but K_m will remain the same as the actual binding of the substrate, by definition, will still function properly.

- In mixed inhibition, the inhibitor can bind to the enzyme at the same time as the enzyme's substrate. However, the binding of the inhibitor affects the binding of the substrate, and vice versa. This type of inhibition can be reduced, but not overcome by increasing concentrations of substrate. Although it is possible for mixed-type inhibitors to bind in the active site, this type of inhibition generally results from an allosteric effect where the inhibitor binds to a different site on an enzyme. Inhibitor binding to this allosteric site changes the conformation (i.e., tertiary structure or three-dimensional shape) of the enzyme so that the affinity of the substrate for the active site is reduced.

Quantitative Description of Reversible Inhibition

• Competitive inhibitors can bind to E, but not to ES. Competitive inhibition increases K_m (i.e., the inhibitor interferes with substrate binding), but does not affect V_{max} (the inhibitor does not hamper catalysis in ES because it cannot bind to ES). • Non-competitive inhibitors have identical affinities for E and ES ($K_i = K_i'$). Non-competitive inhibition does not change K_m (i.e., it does not affect substrate binding) but decreases V_{max} (i.e., inhibitor binding hampers catalysis). • Mixed-type inhibitors bind to both E and ES, but their affinities for these two forms of the enzyme are different ($K_i \neq K_i'$). Thus, mixed-type inhibitors interfere with substrate binding (increase K_m) and hamper catalysis in the ES complex (decrease V_{max}).	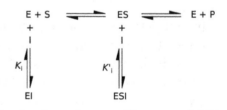 Kinetic scheme for reversible enzyme inhibitors

Reversible inhibition can be described quantitatively in terms of the inhibitor's binding to the enzyme and to the enzyme-substrate complex, and its effects on the kinetic constants of the enzyme. In the classic Michaelis-Menten scheme below, an enzyme (E) binds to its substrate (S) to form the enzyme–substrate complex ES. Upon catalysis, this complex breaks down to release product P and free enzyme. The inhibitor (I) can bind to either E or ES with the dissociation constants K_i or K_i', respectively.

When an enzyme has multiple substrates, inhibitors can show different types of inhibition depending on which substrate is considered. This results from the active site containing two different binding sites within the active site, one for each substrate. For example, an inhibitor might compete with substrate A for the first binding site, but be a non-competitive inhibitor with respect to substrate B in the second binding site.

Measuring The Dissociation Constants of A Reversible Inhibitor

As noted above, an enzyme inhibitor is characterised by its two dissociation constants, K_i and K_i', to the enzyme and to the enzyme-substrate complex, respectively. The enzyme-inhibitor constant K_i can be measured directly by various methods; one extremely accurate method is isothermal titration calorimetry, in which the inhibitor is titrated into a solution of enzyme and the heat released or absorbed is measured. However, the other dissociation constant K_i' is difficult to measure directly, since the enzyme-substrate complex is short-lived and undergoing a chemical reaction to form the product. Hence, K_i' is usually measured indirectly, by observing the enzyme activity under various substrate and inhibitor concentrations, and fitting the data to a modified Michaelis–Menten equation

$$V = \frac{V_{max}[S]}{\alpha K_m + \alpha'[S]} = \frac{(1/\alpha')V_{max}[S]}{(\alpha/\alpha')K_m + [S]}$$

where the modifying factors α and α' are defined by the inhibitor concentration and its two dissociation constants

$$\alpha = 1 + \frac{[I]}{K_i}$$

$$\alpha' = 1 + \frac{[I]}{K_i'}.$$

Thus, in the presence of the inhibitor, the enzyme's effective K_m and V_{max} become $(\alpha/\alpha')K_m$ and $(1/\alpha')V_{max}$, respectively. However, the modified Michaelis-Menten equation as-sumes that binding of the inhibitor to the enzyme has reached equilibrium, which may be a very slow process for inhibitors with sub-nanomolar dissociation constants. In these cases, it is usually more practical to treat the tight-binding inhibitor as an irreversible inhibitor; however, it can still be possible to estimate K_i' kinetically if K_i is measured independently.

The effects of different types of reversible enzyme inhibitors on enzymatic activity can be visualized using graphical representations of the Michaelis–Menten equation, such as Lineweaver–Burk and Eadie-Hofstee plots. For example, in the Lineweaver–Burk plots at the right, the competitive inhibition lines intersect on the y-axis, illustrating that such inhibitors do not affect V_{max}. Similarly, the non-competitive inhibition lines intersect on the x-axis, showing these inhibitors do not affect K_m. However, it can be difficult to estimate K_i and K_i' accurately from such plots, so it is advisable to estimate these constants using more reliable nonlinear regression methods, as described above.

Reversible Inhibitors

Traditionally reversible enzyme inhibitors have been classified as competitive, uncompetitive, or non-competitive, according to their effects on K_m and V_{max}. These different effects result from the inhibitor binding to the enzyme E, to the enzyme–substrate complex ES, or to both, respectively. The division of these classes arises from a problem in their derivation and results in the need to use two different binding constants for one binding event. The binding of an inhibitor and its effect on the enzymatic activity are two distinctly different things, another problem the traditional equations fail to acknowledge. In noncompetitive inhibition the binding of the inhibitor results in 100% inhibition of the enzyme only, and fails to consider the possibility of anything in between. The common form of the inhibitory term also obscures the relationship between the inhibitor binding to the enzyme and its relationship to any other binding term be it the Michaelis–Menten equation or a dose response curve associated with ligand receptor binding. To demonstrate the relationship the following rearrangement can be made:

$$\frac{V_{max}}{1+\frac{[I]}{K_i}} = \frac{V_{max}}{\frac{[I]+K_i}{K_i}}$$

Adding zero to the bottom ([I]-[I])

$$\frac{\frac{V_{max}}{[I]+K_i}}{[I]+K_i-[I]}$$

Dividing by [I]+K_i

$$\frac{\frac{V_{max}}{1}}{1-\frac{[I]}{[I]+K_i}} = V_{max} - V_{max}\frac{[I]}{[I]+K_i}$$

This notation demonstrates that similar to the Michaelis–Menten equation, where the rate of reaction depends on the percent of the enzyme population interacting with substrate.

fraction of the enzyme population bound by substrate

$$\frac{[S]}{[S]+K_m}$$

fraction of the enzyme population bound by inhibitor

$$\frac{[I]}{[I]+K_i}$$

the effect of the inhibitor is a result of the percent of the enzyme population interacting with inhibitor. The only problem with this equation in its present form is that it assumes absolute inhibition of the enzyme with inhibitor binding, when in fact there can be a wide range of effects anywhere from 100% inhibition of substrate turn over to just >0%. To account for this the equation can be easily modified to allow for different degrees of inhibition by including a delta V_{max} term.

$$V_{max} - \Delta V_{max} \frac{[I]}{[I]+K_i}$$

or

$$V_{max1} - (V_{max1} - V_{max2}) \frac{[I]}{[I]+K_i}$$

This term can then define the residual enzymatic activity present when the inhibitor is interacting with individual enzymes in the population. However the inclusion of this term has the added value of allowing for the possibility of activation if the secondary V_{max} term turns out to be higher than the initial term. To account for the possibly of activation as well the notation can then be rewritten replacing the inhibitor "I" with a modifier term denoted here as "X".

$$V_{max1} - (V_{max1} - V_{max2}) \frac{[X]}{[X]+K_x}$$

While this terminology results in a simplified way of dealing with kinetic effects relating to the maximum velocity of the Michaelis–Menten equation, it highlights potential problems with the term used to describe effects relating to the K_m. The K_m relating to the affinity of the enzyme for the substrate should in most cases relate to potential changes in the binding site of the enzyme which would directly result from enzyme inhibitor interactions. As such a term similar to the one proposed above to modulate V_{max} should be appropriate in most situations:

$$K_{m1} - (K_{m1} - K_{m2}) \frac{[X]}{[X]+K_x}$$

Special Cases

- The mechanism of partially competitive inhibition is similar to that of non-competitive, except that the EIS complex has catalytic activity, which may be lower or even higher (partially competitive activation) than that of the enzyme–substrate (ES) complex. This inhibition typically displays a lower V_{max}, but an unaffected K_m value.

- Uncompetitive inhibition occurs when the inhibitor binds only to the enzyme–substrate complex, not to the free enzyme; the EIS complex is catalytically inactive. This mode of inhibition is rare and causes a decrease in both V_{max} and the

K_m value.

- Substrate and product inhibition is where either the substrate or product of an enzyme reaction inhibit the enzyme's activity. This inhibition may follow the competitive, uncompetitive or mixed patterns. In substrate inhibition there is a progressive decrease in activity at high substrate concentrations. This may indicate the existence of two substrate-binding sites in the enzyme. At low substrate, the high-affinity site is occupied and normal kinetics are followed. However, at higher concentrations, the second inhibitory site becomes occupied, inhibiting the enzyme. Product inhibition is often a regulatory feature in metabolism and can be a form of negative feedback.

- Slow-tight inhibition occurs when the initial enzyme–inhibitor complex EI undergoes isomerisation to a second more tightly held complex, EI*, but the overall inhibition process is reversible. This manifests itself as slowly increasing enzyme inhibition. Under these conditions, traditional Michaelis–Menten kinetics give a false value for K_i, which is time–dependent. The true value of K_i can be obtained through more complex analysis of the on (k_{on}) and off (k_{off}) rate constants for inhibitor association. See irreversible inhibition below for more information.

Examples of Reversible Inhibitors

As enzymes have evolved to bind their substrates tightly, and most reversible inhibitors bind in the active site of enzymes, it is unsurprising that some of these inhibitors are strikingly similar in structure to the substrates of their targets. An example of these substrate mimics are the protease inhibitors, a very successful class of antiretroviral drugs used to treat HIV. The structure of ritonavir, a protease inhibitor based on a peptide and containing three peptide bonds, is shown on the right. As this drug resembles the protein that is the substrate of the HIV protease, it competes with this substrate in the enzyme's active site.

Peptide-based HIV-1 protease inhibitor ritonavir

Enzyme inhibitors are often designed to mimic the transition state or intermediate of an enzyme-catalyzed reaction. This ensures that the inhibitor exploits the transition state stabilising effect of the enzyme, resulting in a better binding affinity (lower K_i) than substrate-based designs. An example of such a transition state inhibitor is the antiviral drug oseltamivir; this drug mimics the planar nature of the ring oxonium ion in the reaction of the viral enzyme neuraminidase.

However, not all inhibitors are based on the structures of substrates. For example, the structure of another HIV protease inhibitor tipranavir is shown on the left. This molecule is not based on a peptide and has no obvious structural similarity to a protein substrate. These non-peptide inhibitors can be more stable than inhibitors containing peptide bonds, because they will not be substrates for peptidases and are less likely to be degraded.

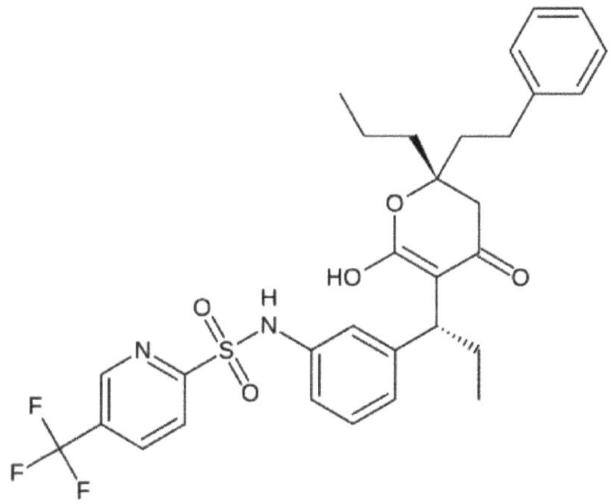

Nonpeptidic HIV-1 protease inhibitor tipranavir

In drug design it is important to consider the concentrations of substrates to which the target enzymes are exposed. For example, some protein kinase inhibitors have chemical structures that are similar to adenosine triphosphate, one of the substrates of these enzymes. However, drugs that are simple competitive inhibitors will have to compete with the high concentrations of ATP in the cell. Protein kinases can also be inhibited by competition at the binding sites where the kinases interact with their substrate proteins, and most proteins are present inside cells at concentrations much lower than the concentration of ATP. As a consequence, if two protein kinase inhibitors both bind in the active site with similar affinity, but only one has to compete with ATP, then the competitive inhibitor at the protein-binding site will inhibit the enzyme more effectively.

Irreversible Inhibitors

Types of Irreversible Inhibition

Irreversible inhibitors usually covalently modify an enzyme, and inhibition can therefore not be reversed. Irreversible inhibitors often contain reactive functional groups such as nitrogen mustards, aldehydes, haloalkanes, alkenes, Michael acceptors, phenyl sulfonates, or fluorophosphonates. These electrophilic groups react with amino acid side chains to form covalent adducts. The residues modified are those with side chains containing nucleophiles such as hydroxyl or sulfhydryl groups; these include the amino acids serine (as in DFP, right), cysteine, threonine, or tyrosine.

Irreversible inhibition is different from irreversible enzyme inactivation. Irreversible inhibitors are generally specific for one class of enzyme and do not inactivate all proteins; they do not function by destroying protein structure but by specifically altering the active site of their target. For example, extremes of pH or temperature usually cause denaturation of all protein structure, but this is a non-specific effect. Similarly, some non-specific chemical treatments destroy protein structure: for example, heating in concentrated hydrochloric acid will hydrolyse the peptide bonds holding proteins together, releasing free amino acids.

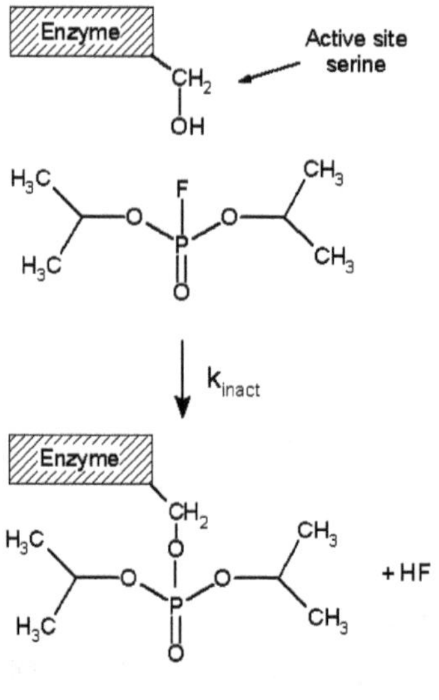

Reaction of the irreversible inhibitor diisopropylfluorophosphate (DFP) with a serine protease

Irreversible inhibitors display time-dependent inhibition and their potency therefore cannot be characterised by an IC_{50} value. This is because the amount of active enzyme at a given concentration of irreversible inhibitor will be different depending on how long the inhibitor is pre-incubated with the enzyme. Instead, $k_{obs}/[I]$ values are used, where k_{obs} is the observed pseudo-first order rate of inactivation (obtained by plotting the log of % activity vs. time) and $[I]$ is the concentration of inhibitor. The $k_{obs}/[I]$ parameter is valid as long as the inhibitor does not saturate binding with the enzyme (in which case $k_{obs} = k_{inact}$).

Analysis of Irreversible Inhibition

As shown in the figure to the left, irreversible inhibitors form a reversible non-covalent complex with the enzyme (EI or ESI) and this then reacts to produce the covalently modified "dead-end complex" EI*. The rate at which EI* is formed is called the inactivation rate or k_{inact}. Since formation of EI may compete with ES, binding of irreversible inhibitors can be prevented by competition either with substrate or with a second, reversible inhibitor. This protection effect is good evidence of a specific reaction of the irreversible inhibitor with the active site.

The binding and inactivation steps of this reaction are investigated by incubating the enzyme with inhibitor and assaying the amount of activity remaining over time. The activity will be decreased in a time-dependent manner, usually following exponential decay. Fitting these data to a rate equation gives the rate of inactivation at this concentration of inhibitor. This is done at several different concentrations of inhibitor. If a reversible EI complex is involved the inactivation rate will be saturable and fitting this curve will give k_{inact} and K_i.

$$E + S \rightleftharpoons ES \rightleftharpoons E + P$$

Kinetic scheme for irreversible inhibitors

Another method that is widely used in these analyses is mass spectrometry. Here, accurate measurement of the mass of the unmodified native enzyme and the inactivated enzyme gives the increase in mass caused by reaction with the inhibitor and shows the stoichiometry of the reaction. This is usually done using a MALDI-TOF mass spectrometer. In a complementary technique, peptide mass fingerprinting involves digestion of the native and modified protein with a protease such as trypsin. This will produce a set of peptides that can be analysed using a mass spectrometer. The peptide that changes in mass after reaction with the inhibitor will be the one that contains the site of modification.

Special Cases

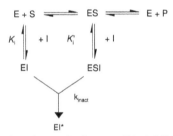

Chemical mechanism for irreversible inhibition of ornithine decarboxylase by DFMO. Pyridoxal 5'-phosphate (Py) and enzyme (E) are not shown. Adapted from

Not all irreversible inhibitors form covalent adducts with their enzyme targets. Some reversible inhibitors bind so tightly to their target enzyme that they are essentially irreversible. These tight-binding inhibitors may show kinetics similar to covalent irreversible inhibitors. In these cases, some of these inhibitors rapidly bind to the enzyme in a low-affinity EI complex and this then undergoes a slower rearrangement to a very tightly bound EI* complex. This kinetic behaviour is called slow-binding. This slow rearrangement after binding often involves a conformational change as the enzyme "clamps down" around the inhibitor molecule. Examples of slow-binding inhibitors include some important drugs, such methotrexate, allopurinol, and the activated form of acyclovir.

Examples of Irreversible Inhibitors

Diisopropylfluorophosphate (DFP) is shown as an example of an irreversible protease inhibitor in the figure above right. The enzyme hydrolyses the phosphorus–fluorine bond, but the phosphate residue remains bound to the serine in the active site, deactivating it. Similarly, DFP also reacts with the active site of acetylcholine esterase in the synapses of neurons, and consequently is a potent neurotoxin, with a lethal dose of less than 100 mg.

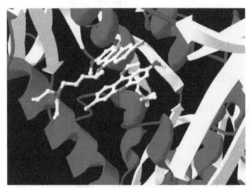

Trypanothione reductase with the lower molecule of an inhibitor bound irreversibly and the upper one reversibly. Created from PDB 1GXF.

Suicide inhibition is an unusual type of irreversible inhibition where the enzyme converts the inhibitor into a reactive form in its active site. An example is the inhibitor of polyamine biosynthesis, α-difluoromethylornithine or DFMO, which is an analogue of the amino acid ornithine, and is used to treat African trypanosomiasis (sleeping sickness). Ornithine decarboxylase can catalyse the decarboxylation of DFMO instead of ornithine, as shown above. However, this decarboxylation reaction is followed by the elimination of a fluorine atom, which converts this catalytic intermediate into a conjugated imine, a highly electrophilic species. This reactive form of DFMO then reacts with either a cysteine or lysine residue in the active site to irreversibly inactivate the enzyme.

Since irreversible inhibition often involves the initial formation of a non-covalent EI complex, it is sometimes possible for an inhibitor to bind to an enzyme in more than one way. For example, in the figure showing trypanothione reductase from the human protozoan parasite *Trypanosoma cruzi*, two molecules of an inhibitor called *quinacrine mustard* are bound in its active site. The top molecule is bound reversibly, but the lower one is bound covalently as it has reacted with an amino acid residue through its nitrogen mustard group.

Discovery and Design of Inhibitors

New drugs are the products of a long drug development process, the first step of which is often the discovery of a new enzyme inhibitor. In the past the only way to discover these new inhibitors was by trial and error: screening huge libraries of compounds against a target enzyme and hoping that some useful leads would emerge. This brute force approach is still successful and has even been extended by combinatorial chemistry approaches that quickly produce large numbers of novel compounds and high-throughput screening technology to rapidly screen these huge chemical libraries for useful inhibitors.

Robots used for the high-throughput screening of chemical libraries to discover new enzyme inhibitors

More recently, an alternative approach has been applied: rational drug design uses the three-dimensional structure of an enzyme's active site to predict which molecules might be inhibitors. These predictions are then tested and one of these tested compounds may be a novel inhibitor. This new inhibitor is then used to try to obtain a structure of the enzyme in an inhibitor/enzyme complex to show how the molecule is binding to the active site, allowing changes to be made to the inhibitor to try to optimise binding. This test and improve cycle is then repeated until a sufficiently potent inhibitor is produced. Computer-based methods of predicting the affinity of an inhibitor for an enzyme are also being developed, such as molecular docking and molecular mechanics.

Uses of Inhibitors

Enzyme inhibitors are found in nature and are also designed and produced as part of pharmacology and biochemistry. Natural poisons are often enzyme inhibitors that have evolved to defend a plant or animal against predators. These natural toxins include some of the most poisonous compounds known. Artificial inhibitors are often used as drugs, but can also be insecticides such as malathion, herbicides such as glyphosate, or disinfectants such as triclosan. Other artificial enzyme inhibitors block acetylcholinesterase, an enzyme which breaks down acetylcholine, and are used as nerve agents in chemical warfare.

Chemotherapy

An example of a medicinal enzyme inhibitor is sildenafil (Viagra), a common treatment for male erectile dysfunction. This compound is a potent inhibitor of cGMP specific phosphodiesterase type 5, the enzyme that degrades the signalling molecule cyclic guanosine monophosphate. This signalling molecule triggers smooth muscle relaxation and allows blood flow into the corpus cavernosum, which causes an erection. Since the drug decreases the activity of the enzyme that halts the signal, it makes this signal last for a longer period of time.

The structure of sildenafil (Viagra)

The coenzyme folic acid (left) compared to the anti-cancer drug methotrexate (right)

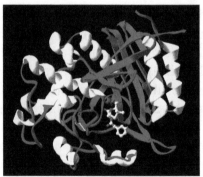

The structure of a complex between penicillin G and the *Streptomyces* transpeptidase. Generated from PDB 1PWC. The most common uses for enzyme inhibitors are as drugs to treat disease. Many of these inhibitors target a human enzyme and aim to correct a pathological condition. However, not all drugs are enzyme inhibitors. Some, such as anti-epileptic drugs, alter enzyme activity by causing more or less of the enzyme to be produced. These effects are called enzyme induction and inhibition and are alterations in gene expression, which is unrelated to the type of enzyme inhibition discussed here. Other drugs interact with cellular targets that are not enzymes, such as ion channels or membrane receptors.

Another example of the structural similarity of some inhibitors to the substrates of the enzymes they target is seen in the figure comparing the drug methotrexate to folic acid. Folic acid is a substrate of dihydrofolate reductase, an enzyme involved in making nucleotides that is potently inhibited by methotrexate. Methotrexate blocks the action of dihydrofolate reductase and thereby halts the production of nucleotides. This block of nucleotide biosynthesis is more toxic to rapidly growing cells than non-dividing cells, since a rapidly growing cell has to carry out DNA replication, therefore methotrexate is often used in cancer chemotherapy.

Antibiotics

Drugs also are used to inhibit enzymes needed for the survival of pathogens. For example, bacteria are surrounded by a thick cell wall made of a net-like polymer called peptidoglycan. Many antibiotics such as penicillin and vancomycin inhibit the enzymes that produce and then cross-link the strands of this polymer together. This causes the cell wall to lose strength and the bacteria to burst. In the figure, a molecule of penicillin (shown in a ball-and-stick form) is shown bound to its target, the transpeptidase from the bacteria *Streptomyces* R61 (the protein is shown as a ribbon-diagram).

Antibiotic drug design is facilitated when an enzyme that is essential to the pathogen's survival is absent or very different in humans. In the example above, humans do not make peptidoglycan, therefore inhibitors of this process are selectively toxic to bacteria. Selective toxicity is also produced in antibiotics by exploiting differences in the structure of the ribosomes in bacteria, or how they make fatty acids.

Metabolic Control

Enzyme inhibitors are also important in metabolic control. Many metabolic pathways in the cell are inhibited by metabolites that control enzyme activity through allosteric regulation or substrate inhibition. A good example is the allosteric regulation of the glycolytic pathway. This catabolic pathway consumes glucose and produces ATP, NADH and pyruvate. A key step for the regulation of glycolysis is an early reaction in the pathway catalysed by phosphofructokinase-1 (PFK1). When ATP levels rise, ATP binds an allosteric site in PFK1 to decrease the rate of the enzyme reaction; glycolysis is inhibited and ATP production falls. This negative feedback control helps maintain a steady concentration of ATP in the cell. However, metabolic pathways are not just regulated through inhibition since enzyme activation is equally important. With respect to PFK1, fructose 2,6-bisphosphate and ADP are examples of metabolites that are allosteric activators.

Physiological enzyme inhibition can also be produced by specific protein inhibitors. This mechanism occurs in the pancreas, which synthesises many digestive precursor enzymes known as zymogens. Many of these are activated by the trypsin protease, so it is important to inhibit the activity of trypsin in the pancreas to prevent the organ from digesting itself. One way in which the activity of trypsin is controlled is the production of a specific and potent trypsin inhibitor protein in the pancreas. This inhibitor binds tightly to trypsin, preventing the trypsin activity that would otherwise be detrimental to the organ. Although the trypsin inhibitor is a protein, it avoids being hydrolysed as a substrate by the protease by excluding water from trypsin's active site and destabilising the transition state. Other examples of physiological enzyme inhibitor proteins include the barstar inhibitor of the bacterial ribonuclease barnase and the inhibitors of protein phosphatases.

Pesticides

Many pesticides are enzyme inhibitors. Acetylcholinesterase (AChE) is an enzyme found in animals from insects to humans. It is essential to nerve cell function through its mechanism of breaking down the neurotransmitter acetylcholine into its constituents, acetate and choline. This is somewhat unique among neurotransmitters as most, including serotonin, dopamine, and norepinephrine, are absorbed from the synaptic cleft rather than cleaved. A large number of AChE inhibitors are used in both medicine and agriculture. Reversible competitive inhibitors, such as edrophonium, physostigmine, and neostigmine, are used in the treatment of myasthenia gravis and in anaesthesia. The carbamate pesticides are also examples of reversible AChE inhibitors. The organophosphate pesticides such as malathion, parathion, and chlorpyrifos irreversibly inhibit acetylcholinesterase.

The herbicide glyphosate is an inhibitor of 3-phosphoshikimate 1-carboxyvinyltransferase, other herbicides, such as the sulfonylureas inhibit the enzyme acetolactate synthase. Both these enzymes are needed for plants to make branched-chain amino acids. Many other enzymes are inhibited by herbicides, including enzymes needed for the biosynthesis of lipids and carotenoids and the processes of photosynthesis and oxidative phosphorylation.

To discourage seed predators, pulses contain trypsin inhibitors that interfere with digestion.

Natural Poisons

Animals and plants have evolved to synthesise a vast array of poisonous products including secondary metabolites, peptides and proteins that can act as inhibitors. Natural toxins are usually small organic molecules and are so diverse that there are probably natural inhibitors for most metabolic processes. The metabolic processes targeted by natural poisons encompass more than enzymes in metabolic pathways and can also include the inhibition of receptor, channel and structural protein functions in a cell. For example, paclitaxel (taxol), an organic molecule found in the Pacific yew tree, binds tightly to tubulin dimers and inhibits their assembly into microtubules in the cytoskeleton.

Many natural poisons act as neurotoxins that can cause paralysis leading to death and have functions for defence against predators or in hunting and capturing prey. Some of these natural inhibitors, despite their toxic attributes, are valuable for therapeutic uses at lower doses. An example of a neurotoxin are the glycoalkaloids, from the plant species in the *Solanaceae* family (includes potato, tomato and eggplant), that are acetylcholinesterase inhibitors. Inhibition of this enzyme causes an uncontrolled increase in the acetylcholine neurotransmitter, muscular paralysis and then death. Neurotoxicity can also result from the inhibition of receptors; for example, atropine from deadly nightshade (*Atropa belladonna*) that functions as a competitive antagonist of the muscarinic acetylcholine receptors.

Although many natural toxins are secondary metabolites, these poisons also include peptides and proteins. An example of a toxic peptide is alpha-amanitin, which is found in relatives of the death cap mushroom. This is a potent enzyme inhibitor, in this case preventing the RNA polymerase II enzyme from transcribing DNA. The algal toxin microcystin is also a peptide and is an inhibitor of protein phosphatases. This toxin can contaminate water supplies after algal blooms and is a known carcinogen that can also cause acute liver hemorrhage and death at higher doses.

Proteins can also be natural poisons or antinutrients, such as the trypsin inhibitors (discussed above) that are found in some legumes, as shown in the figure above. A less common class of toxins are toxic enzymes: these act as irreversible inhibitors of their target enzymes and work by chemically modifying their substrate enzymes. An example is ricin, an extremely potent protein toxin found in castor oil beans. This enzyme is a glycosidase that inactivates ribosomes. Since ricin is a catalytic irreversible inhibitor, this allows just a single molecule of ricin to kill a cell.

Forms of Enzyme Inhibition

Competitive Inhibition

Competitive inhibition is a form of enzyme inhibition where binding of the inhibitor to the active site on the enzyme prevents binding of the substrate and *vice versa*.

Most competitive inhibitors function by binding reversibly to the active site of the enzyme. As a result, many sources state that this is the defining feature of competitive inhibitors. This, however, is a misleading oversimplification, as there are many possible mechanisms by which an enzyme may bind either the inhibitor or the substrate but never both at the same time. For example, allosteric inhibitors may display competitive, non-competitive, or uncompetitive inhibition.

Mechanism

In competitive inhibition, at any given moment, the enzyme may be bound to the inhibitor, the substrate, or neither, but it cannot bind both at the same time.

In virtually every case, competitive inhibitors bind in the same binding site as the substrate, but same-site binding is not a requirement. A competitive inhibitor could bind to an allosteric site of the free enzyme and prevent substrate binding, as long as it does not bind to the allosteric site when the substrate is bound. For example, strychnine acts as an allosteric inhibitor of the glycine receptor in the mammalian spinal cord and brain stem. Glycine is a major post-synaptic inhibitory neurotransmitter with a specific receptor site. Strychnine binds to an alternate site that reduces the affinity of the glycine receptor for glycine, resulting in convulsions due to lessened inhibition by the glycine.

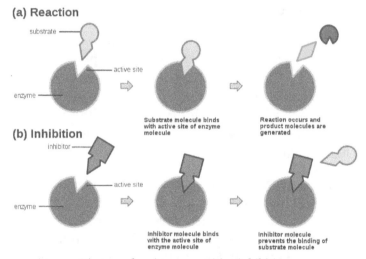

Diagram showing competitive inhibition

In competitive inhibition, the maximum velocity (V_{max}) of the reaction is unchanged, while the apparent affinity of the substrate to the binding site is decreased (the K_d dissociation constant is apparently increased). The change in K_m (Michaelis-Menten constant) is parallel to the alteration in K_d. Any given competitive inhibitor concentration can be overcome by increasing the substrate concentration in which case the substrate will outcompete the inhibitor in binding to the enzyme.

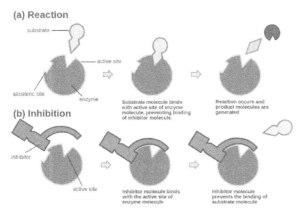

Competitive inhibition can also be allosteric, as long as the inhibitor and the substrate cannot bind the enzyme at the same time.

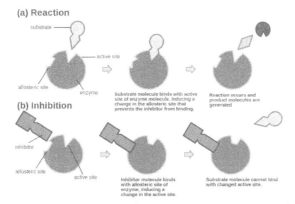

Another possible mechanism for allosteric competitive inhibition.

Equation

Competitive inhibition increases the apparent value of the Michaelis-Menten constant, K_m^{app}, such that initial rate of reaction, V_0, is given by

$$V_0 = \frac{V_{max}[S]}{K_m^{app} + [S]}$$

where $K_m^{app} = K_m(1+[I]/K_i)$, K_i is the inhibitor's dissociation constant and $[I]$ is the inhibitor concentration.

V_{max} remains the same because the presence of the inhibitor can be overcome by higher substrate concentrations. K_m^{app}, the substrate concentration that is needed to reach $V_{max}/2$, increases with the presence of a competitive inhibitor. This is because the concentration of substrate needed to reach with V_{max} an inhibitor is greater than the concentration of substrate needed to reach V_{max} without an inhibitor.

Derivation

In the simplest case of a single-substrate enzyme obeying Michaelis-Menten kinetics, the typical scheme

$$E + S \underset{k_{-1}}{\overset{k_1}{\rightleftharpoons}} ES \overset{k_2}{\longrightarrow} E + P$$

is modified to include binding of the inhibitor to the free enzyme:

$$EI + S \underset{k_3}{\overset{k_{-3}}{\rightleftharpoons}} E + S + I \underset{k_{-1}}{\overset{k_1}{\rightleftharpoons}} ES + I \overset{k_2}{\longrightarrow} E + P + I$$

Note that the inhibitor does not bind to the ES complex and the substrate does not bind to the EI complex. It is generally assumed that this behavior is indicative of both compounds binding at the same site, but that is not strictly necessary. As with the derivation of the Michaelis-Menten equation, assume that the system is at steady-state, i.e. the concentration of each of the enzyme species is not changing.

$$\frac{d[E]}{dt} = \frac{d[ES]}{dt} = \frac{d[EI]}{dt} = 0.$$

Furthermore, the known total enzyme concentration is , and the velocity is measured under conditions in which the substrate and inhibitor concentrations do not change substantially and an insignificant amount of product has accumulated.

We can therefore set up a system of equations:

$$[E]_0 = [E] + [ES] + [EI] \tag{1}$$

$$\frac{d[E]}{dt} = 0 = -k_1[E][S] + k_{-1}[ES] + k_2[ES] - k_3[E][I] + k_{-3}[EI] \tag{2}$$

$$\frac{d[ES]}{dt} = 0 = k_1[E][S] - k_{-1}[ES] - k_2[ES] \tag{3}$$

$$\frac{d[EI]}{dt} = 0 = k_3[E][I] - k_{-3}[EI] \tag{4}$$

where $[S], [I]$ and $[E]_0$ are known. The initial velocity is defined as $V_0 = d[P]/dt = k_2[ES]$, so we need to define the unknown $[ES]$ in terms of the knowns $[S], [I]$ and $[E]_0$.

From equation (3), we can define E in terms of ES by rearranging to

$$k_1[E][S] = (k_{-1} + k_2)[ES]$$

Dividing by $k_1[S]$ gives

$$[E] = \frac{(k_{-1} + k_2)[ES]}{k_1[S]}$$

As in the derivation of the Michaelis-Menten equation, the term $(k_{-1} + k_2)/k_1$ can be replaced by the macroscopic rate constant K_m:

$$[E] = \frac{K_m[ES]}{[S]} \tag{5}$$

Substituting equation (5) into equation (4), we have

$$0 = \frac{k_3[I]K_m[ES]}{[S]} - k_{-3}[EI]$$

Rearranging, we find that

$$[EI] = \frac{K_m k_3[I][ES]}{k_{-3}[S]}$$

At this point, we can define the dissociation constant for the inhibitor as $K_i = k_{-3}/k_3$, giving

$$[EI] = \frac{K_m[I][ES]}{K_i[S]} \tag{6}$$

At this point, substitute equation (5) and equation (6) into equation (1):

$$[E]_0 = \frac{K_m[ES]}{[S]} + [ES] + \frac{K_m[I][ES]}{K_i[S]}$$

Rearranging to solve for ES, we find

$$[E]_0 = [ES]\left(\frac{K_m}{[S]} + 1 + \frac{K_m[I]}{K_i[S]}\right) = [ES]\frac{K_m K_i + K_i[S] + K_m[I]}{K_i[S]}$$

$$[ES] = \frac{K_i[S][E]_0}{K_m K_i + K_i[S] + K_m[I]} \tag{7}$$

Returning to our expression for V_0, we now have:

$$V_0 = k_2[ES] = \frac{k_2 K_i[S][E]_0}{K_m K_i + K_i[S] + K_m[I]}$$

$$V_0 = \frac{k_2[E]_0[S]}{K_m + [S] + K_m \frac{[I]}{K_i}}$$

Since the velocity is maximal when all the enzyme is bound as the enzyme-substrate complex, $V_{max} = k_2[E]_0$. Replacing and combining terms finally yields the conventional form:

$$V_0 = \frac{V_{max}[S]}{K_m\left(1 + \frac{[I]}{K_i}\right) + [S]} \tag{8}$$

To compute the concentration of competitive inhibitor [I] that yields a fraction f_{V_0} of velocity V_0 V_0 where $0 < f_{V_0} < 1$:

$$[I] = (1/fV0 - 1)K_i\left(1 + \frac{[S]}{K_m}\right) \tag{9}$$

Non-Competitive Inhibition

Non-competitive inhibition is a type of enzyme inhibition where the inhibitor reduces the activity of the enzyme and binds equally well to the enzyme whether or not it has already bound the substrate.

The inhibitor may bind to the enzyme whether or not the substrate has already been bound, but if it has a higher affinity for binding the enzyme in one state or the other, it is called a mixed inhibitor.

Terminology

It is important to note that while all non-competitive inhibitors bind the enzyme at allosteric sites (i.e. locations other than its active site)—not all inhibitors that bind at allosteric sites are non-competitive inhibitors. In fact, allosteric inhibitors may act as competitive, non-competitive, or uncompetitive inhibitors.

Many sources continue to conflate these two terms, or state the definition of allosteric inhibition as the definition for non-competitive inhibition.

Mechanism

Non-competitive inhibition models a system where the inhibitor and the substrate may both be bound to the enzyme at any given time. When both the substrate and the inhibitor are bound, the enzyme-substrate-inhibitor complex cannot form product and can only be converted back to the enzyme-substrate complex or the enzyme-inhibitor complex. Non-competitive inhibition is distinguished from general mixed inhibition in that the inhibitor has an equal affinity for the enzyme and the enzyme-substrate complex.

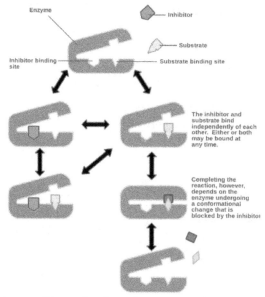

Illustration of a possible mechanism of non-competitive or mixed inhibition.

The most common mechanism of non-competitive inhibition involves reversible binding of the inhibitor to an allosteric site, but it is possible for the inhibitor to operate via other means including direct binding to the active site. It differs from competitive inhibition in that the binding of the inhibitor does not prevent binding of substrate, and vice versa, it simply prevents product formation for a limited time.

This type of inhibition reduces the maximum rate of a chemical reaction without changing the apparent binding affinity of the catalyst for the substrate (K_m^{app} – see Michaelis-Menten kinetics).

Equation

In the presence of a non-competitive inhibitor, the apparent enzyme affinity is equivalent to the actual affinity. In terms of Michaelis-Menten kinetics, $K_m^{app} = K_m$. This can be seen as a consequence of Le Chatelier's principle because the inhibitor binds to both the enzyme and the enzyme-substrate complex equally so that the equilibrium is maintained. However, since some enzyme is always inhibited from converting the substrate to product, the effective enzyme concentration is lowered.

Mathematically,

$$V_{max}^{app} = \frac{V_{max}}{1 + \dfrac{[I]}{K_I}}$$

$$apparent\,[E]_0 = \frac{[E]_0}{1 + \dfrac{[I]}{K_I}}$$

Example: Noncompetitive Inhibitors of Cyp2c9 Enzyme

Noncompetitive inhibitors of CYP2C9 enzyme include nifedipine, tranylcypromine, phenethyl isothiocyanate, and 6-hydroxyflavone. Computer docking simulation and constructed mutants substituted indicate that the noncompetitive binding site of 6-hydroxyflavone is the reported allosteric binding site of CYP2C9 enzyme.

Uncompetitive Inhibitor

Uncompetitive inhibition, also known as anti-competitive inhibition, takes place when an enzyme inhibitor binds only to the complex formed between the enzyme and the substrate (the E-S complex).

While uncompetitive inhibition requires that an enzyme-substrate complex must be formed, non-competitive inhibition can occur with or without the substrate present.

Mechanism

This reduction in the effective concentration of the E-S complex increases the enzyme's apparent affinity for the substrate through Le Chatelier's principle (K_m is lowered) and decreases the maximum enzyme activity (V_{max}), as it takes longer for the substrate or product to leave the active site. Uncompetitive inhibition works best when substrate concentration is high. An uncompetitive inhibitor need not resemble the substrate of the reaction it is inhibiting.

Mathematical Definition

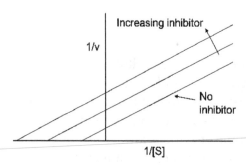

Lineweaver–Burk plot of uncompetitive enzyme inhibition.

The Lineweaver–Burk equation states that:

$$\frac{1}{v} = \frac{K_m}{V_{max}[S]} + \frac{1}{V_{max}}$$

Where v is the initial reaction velocity, K_m is the Michaelis–Menten constant, V_{max} is the maximum reaction velocity, and $[S]$ is the concentration of the substrate.

The Lineweaver–Burk plot for an uncompetitive inhibitor produces a line parallel to the original enzyme-substrate plot, but with a higher y-intercept, due to the presence of an inhibition term $\frac{[I]}{K_i}$:

$$\frac{1}{v} = \frac{K_m}{V_{max}[S]} + \frac{1 + \dfrac{[I]}{K_i}}{V_{max}}$$

Where $[I]$ is the concentration of the inhibitor and K_i is an inhibition constant characteristic of the inhibitor.

Mixed Inhibition

Mixed inhibition is a type of enzyme inhibition in which the inhibitor may bind to the enzyme whether or not the enzyme has already bound the substrate but has a greater affinity for one state or the other. It is called "mixed" because it can be seen as a conceptual "mixture" of competitive inhibition, in which the inhibitor can only bind the enzyme if the substrate *has not* already bound, and uncompetitive inhibition, in which the inhibitor can only bind the enzyme if the substrate *has* already bound. If the ability of the inhibitor to bind the enzyme is *exactly the same* whether or not the enzyme has already bound the substrate, it is known as a non-competitive inhibitor. Non-competitive inhibition is sometimes thought of as a special case of mixed inhibition.

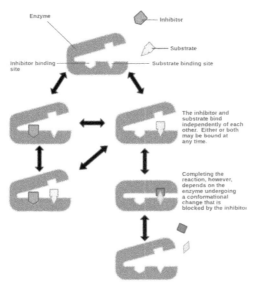

A possible mechanism of non-competitive inhibition, a kind of mixed inhibition.

In mixed inhibition, the inhibitor binds to an allosteric site, i.e. a site different from the active site where the substrate binds. However, not all inhibitors that bind at allosteric sites are mixed inhibitors.

Mixed inhibition may result in either a decrease in the apparent affinity of the enzyme for the substrate ($K_m^{app} > K_m$); a decrease in apparent affinity means the Km value appears to increase) in cases where the inhibitor favors binding the free enzyme, or in an increase in the apparent affinity ($K_m^{app} < K_m$); an increase in apparent affinity means the Km value appears to decrease) when the inhibitor binds favorably to the enzyme-substrate complex. In either case the inhibition decreases the apparent maximum enzyme reaction rate ($V_{max}^{app} < V_{max}$).

Mathematically, mixed inhibition occurs when the factors α and α' (introduced into the Michaelis-Menten equation to account for competitive and uncompetitive inhibition, respectively) are both greater than 1.

In the special case where α = α', noncompetitive inhibition occurs, in which case V_{max}^{app} is reduced but K_m is unaffected. This is very unusual in practice

Enzyme Activator

Enzyme activators are molecules that bind to enzymes and increase their activity. They are the opposite of enzyme inhibitors. These molecules are often involved in the allosteric regulation of enzymes in the control of metabolism. An example of an enzyme activator working in this way is fructose 2,6-bisphosphate, which activates phosphofructokinase 1 and increases the rate of glycolysis in response to the hormone insulin.

In some cases, when a substrate binds to one catalytic subunit of an enzyme, this can trigger an increase in the substrate affinity as well as catalytic activity in the enzyme's other subunits, and thus the substrate acts as an activator.

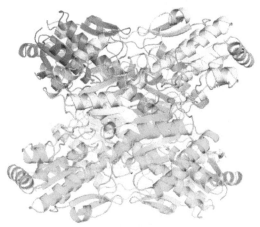

Bacillus stearothermophilus phosphofructokinase. PDB: 6PFK.

Enzyme Promiscuity

Enzyme promiscuity is the ability of an enzyme to catalyse a fortuitous side reaction in addition to its main reaction. Although enzymes are remarkably specific catalysts, they can often perform side reactions in addition to their main, native catalytic activity. These promiscuous activities are usually slow relative to the main activity and are under neutral selection. Despite ordinarily being physiologically irrelevant, under new selective pressures these activities may confer a fitness benefit therefore prompting the evolution of the formerly promiscuous activity to become the new main activity. An example of this is the atrazine chlorohydrolase (*atzA* encoded) from *Pseudomonas sp.* ADP which evolved from melamine deaminase (*triA* encoded), which has very small promiscuous activity towards atrazine, a man-made chemical.

Introduction

Enzymes are evolved to catalyse a particular reaction on a particular substrate with a high catalytic efficiency (k_{cat}/K_M, cf. Michaelis–Menten kinetics). However, in addition to this main activity, they possess other activities that are generally several orders of magnitude lower, and that are not a result of evolutionary selection and therefore do not partake in the physiology of the organism.This phenomenon allows new functions to be gained as the promiscuous activity could confer a fitness benefit under a new selective pressure leading to its duplication and selection as a new main activity.

Enzyme Evolution

Duplication and Divergence

Several theoretical models exist to predict the order of duplication and specialisation events, but the actual process is more intertwined and fuzzy (§ *Reconstructed enzymes* below). On one hand, gene amplification results in an increase in enzyme concentration, and potentially freedom from a restrictive regulation, therefore increasing the reaction rate (v) of the promiscuous activity of the enzyme making its effects more pronounced physiologically ("gene dosage effect"). On the other, enzymes may evolve an increased secondary activity with little loss to the primary activity ("robustness") with little adaptive conflict (§ *Robustness and plasticity* below).

Robustness and Plasticity

A study of three distinct hydrolases (human serum paraoxonase (PON1), pseudomonad phosphotriesterase (PTE) and human carbonic anhydrase II (CAII)) has shown the main activity is "robust" towards change, whereas the promiscuous activities are more "plastic". Specifically, selecting for an activity that is not the main activity (via directed evolution), does not initially diminish the main activity (hence its robustness), but greatly affects the non-selected activities (hence their plasticity).

The phosphotriesterase (PTE) from *Pseudomonas diminuta* was evolved to become an arylesterase (P–O to C–O hydrolase) in eighteen rounds gaining a 10^9 shift in specificity (ratio of K_M), however most of the change occurred in the initial rounds, where the unselected vestigial PTE activity was retained and the evolved arylesterase activity grew, while in the latter rounds there was a little trade-off for the loss of the vestigial PTE activity in favour of the arylesterase activity.

This means firstly that a specialist enzyme (monofunctional) when evolved goes through a generalist stage (multifunctional), before becoming a specialist again —presumably after gene duplication according to the IAD model— and secondly that promiscuous activities are more plastic than the main activity.

Reconstructed Enzymes

The most recent and most clear cut example of enzyme evolution is the rise of bioremediating enzymes in the past 60 years. Due to the very low number of amino acid changes, these provide an excellent model to investigate enzyme evolution in nature. However, using extant enzymes to determine how the family of enzymes evolved has the drawback that the newly evolved enzyme is compared to paralogues without knowing the true identity of the ancestor before the two genes divereged. This issue can be resolved thanks to ancestral reconstruction. First proposed in 1963 by Linus Pauling and Emile Zuckerkandl, ancestral reconstruction is the inference and synthesis of a gene from the ancestral form of a group of genes, which has had a recent revival thanks to improved inference techniques and low-cost artificial gene synthesis, resulting in several ancestral enzymes —dubbed "stemzymes" by some—to be studied.

Evidence gained from reconstructed enzyme suggests that the order of the events where the novel activity is improved and the gene is duplication is not clear cut, unlike what the theoretical models of gene evolution suggest.

One study showed that the ancestral gene of the immune defence protease family in mammals had a broader specificity and a higher catalytic efficiency than the contemporary family of paralogues, whereas another study showed that the ancestral steroid receptor of vertebrates was an oestrogen receptor with slight substrate ambiguity for other hormones —indicating that these probably were not synthesised at the time.

This variability in ancestral specificity has not only been observed between different genes, but also within the same gene family. In light of the large number of paralogous fungal α-glucosidase genes with a number of specific maltose-like (maltose, turanose, maltotriose, maltulose and sucrose) and isomaltose-like (isomaltose and palatinose) substrates, a study reconstructed all key ancestors and found that the last common ancestor of the paralogues was mainly active on maltose-like substrates with only trace activity for isomaltose-like sugars, despite leading to a lineage of iso-maltose glucosidases and a lineage that further split into maltose glucosidases and iso-maltose glucosidases. Antithetically, the ancestor before the latter split had a more pronounced isomaltose-like glucosidase activity.

Primordial Metabolism

Roy Jensen in 1976 theorised that primordial enzymes had to be highly promiscuous in order for metabolic networks to assemble in a patchwork fashion (hence its name, the *patchwork model*). This primordial catalytic versatility was later lost in favour of highly catalytic specialised orthologous enzymes. As a consequence, many central-metabolic enzymes have structural homologues that diverged before the last universal common ancestor.

Distribution

Promiscuity is however not only a primordial trait, in fact it is very widespread property in modern genomes. A series of experiments have been conducted to assess the distribution of promiscuous enzyme activities in *E. coli*. In *E. coli* 21 out of 104 single-gene knockouts tested (from the Keio collection) could be rescued by overexpressing a non-cognate *E. coli* protein (using a pooled set of plasmids of the ASKA collection). The mechanisms by which the noncognate ORF could rescue the knockout can be grouped into eight categories: isozyme overexpression (homologues), substrate ambiguity, transport ambiguity (scavenging), catalytic promiscuity, metabolic flux maintenance (including overexpression of the large component of a synthase in the absence of the amine transferase subunit), pathway bypass, regulatory effects and unknown mechanisms. Similarly, overexpressing the ORF collection allowed *E. coli* to gain over an order of magnitude in resistance in 86 out 237 toxic environment.

Homology

Homologues are sometimes known to display promiscuity towards each other's main reactions. This crosswise promiscuity has been most studied with members of the Alkaline phosphatase superfamily, which catalyse hydrolytic reaction on the sulfate, phosphonate, monophosphate, diphosphate or triphosphate ester bond of several compounds. Despite the divergence the homologues have a varying degree of reciprocal promiscuity: the differences in promiscuity are due to mechanisms involved, particularly the intermediate required.

Degree of Promiscuity

Enzymes are generally in a state that is not only a compromise between stability and catalytic efficiency, but also for specificity and evolvability, the latter two dictating whether an enzyme is a generalist (highly evolvable due to large promiscuity, but low main activity) or a specialist (high main activity, poorly evolvable due to low promiscuity). Examples of these are enzymes for primary and secondary metabolism in plants (§ *Plant secondary metabolism* below). Other factors can come into play, for example the glycerophosphodiesterase (*gpdQ*) from *Enterobacter aerogenes* shows different values for its promiscuous activities depending on the two metal ions it binds, which is dictated by ion availability. In some cases promiscuity can be increased by relaxing the specificity of the active site by enlarging it with a single mutation as was the case of a D297G mutant of the *E. coli* L-Ala-D/L-Glu epimerase (*ycjG*) and E323G mutant of a pseudomonad muconate lactonizing enzyme II, allowing them to promiscuously catalyse the activity of O-succinylbenzoate synthase (*menC*). Conversely, promiscuity can be decreased as was the case of γ-humulene synthase (a sesquiterpene synthase) from *Abies grandis* that is known to produce 52 different sesquiterpenes from farnesyl diphosphate upon several mutations.

Studies on enzymes with broad-specificity—not promiscuous, but conceptually close—such as mammalian trypsin and chymotrypsin, and the bifunctional isopropylmalate isomerase/homoaconitase from *Pyrococcus horikoshii* have revealed that active site loop mobility contributes substantially to the catalytic elasticity of the enzyme.

Toxicity

A promiscuous activity is a non-native activity the enzyme did not evolve to do, but arises due to an accommodating conformation of the active site. However, the main activity of the enzyme is a result not only of selection towards a high catalytic rate towards a particular substrate to produce a particular product, but also to avoid the production of toxic or unnecessary products. For example, if a tRNA synthases loaded an incorrect amino acid onto a tRNA, the resulting peptide would have unexpectedly altered properties, consequently to enhance fidelity several additional domains are present. Similar in reaction to tRNA synthases, the first subunit of tyrocidine synthetase (*tyrA*) from *Bacillus brevis* adenylates a molecule of phenylalanine in order to use the adenyl moiety as a handle to produce tyrocidine, a cyclic non-ribosomal peptide. When the specificity of enzyme was probed, it was found that it was highly selective against natural amino acids that were not phenylalanine, but was much more tolerant towards unnatural amino acids. Specifically, most amino acids were not catalysed, whereas the next most catalysed native amino acid was the structurally similar tyrosine, but at a thousandth as much as phenylalanine, whereas several unnatural amino acids where catalysed better than tyrosine, namely D-phenylalanine, β-cyclohexyl-L-alanine, 4-amino-L-phenylalanine and L-norleucine.

One peculiar case of selected secondary activity are polymerases and restriction endonucleases, where incorrect activity is actually a result of a compromise between fidelity and evolvability. For example, for restriction endonucleases incorrect activity (star activity) is often lethal for the organism, but a small amount allows new functions to evolve against new pathogens.

Plant Secondary Metabolism

Plants produce a large number of secondary metabolites thanks to enzymes that, unlike those involved in primary metabolism, are less catalytically efficient but have a larger mechanistic elasticity (reaction types) and broader specificities. The liberal drift threshold (caused by the low selective pressure due the small population size) allows the fitness gain endowed by one of the products to maintain the other activities even though they may be physiologically useless.

Anthocyanins (delphindin pictured) confer plants, particularly their flowers, with a variety of colours to attract pollinators and a typical example of plant secondary metabolite.

Biocatalysis

In biocatalysis, many reactions are sought that are absent in nature. To do this, enzymes with a small promiscuous activity towards the required reaction are identified and evolved via directed evolution or rational design.

An example of a commonly evolved enzyme is ω-transaminase which can replace a ketone with a chiral amine and consequently libraries of different homologues are commercially available for rapid biomining (*eg.* Codexis).

Another example is the possibility of using the promiscuous activities of cysteine synthase (*cysM*) towards nucleophiles to produce non-proteinogenic amino acids.

Reaction Similarity

Similarity between enzymatic reactions (EC) can be calculated by using bond changes, reaction centres or substructure metrics (EC-BLAST).

Drugs and Promiscuity

Whereas promiscuity is mainly studied in terms of standard enzyme kinetics, drug binding and subsequent reaction is a promiscuous activity as the enzyme catalyses an inactivating reaction towards a novel substrate it did not evolve to catalyse.

Mammalian xenobiotic metabolism, on the other hand, was evolved to have a broad specificity to oxidise, bind and eliminate foreign lipophilic compounds which may be toxic, such as plant alkaloids, so their ability to detoxify anthropogenic xenobiotics is an extension of this.

References

- Irwin H. Segel, Enzyme Kinetics : Behavior and Analysis of Rapid Equilibrium and Steady-State Enzyme Systems. Wiley–Interscience; New edition (1993), ISBN 0-471-30309-7

- Walsh, Ryan (2012). "Ch. 17. Alternative Perspectives of Enzyme Kinetic Modeling". In Ekinci, Deniz. Medicinal Chemistry and Drug Design (PDF). InTech. pp. 357–371. ISBN 978-953-51-0513-8.

- Segel, Irwin H. (1993) Enzyme Kinetics : Behavior and Analysis of Rapid Equilibrium and Steady-State Enzyme Systems. Wiley-Interscience; New edition , ISBN 0-471-30309-7.

- Szedlacsek, SE; Duggleby, RG (1995). "Kinetics of slow and tight-binding inhibitors". Enzyme Kinetics and Mechanism Part D: Developments in Enzyme Dynamics. Methods in Enzymology. 249. pp. 144–80. doi:10.1016/0076-6879(95)49034-5. ISBN 978-0-12-182150-0. PMID 7791610.

- Cohen, J.A.; Oosterbaan, R.A.; Berends, F. (1967). "[81] Organophosphorus compounds". Enzyme Structure. Methods in Enzymology. 11. p. 686. doi:10.1016/S0076-6879(67)11085-9. ISBN 978-0-12-181860-9.

- Dick RM (2011). "Chapter 2. Pharmacodynamics: The Study of Drug Action". In Ouellette R, Joyce JA. Pharmacology for Nurse Anesthesiology. Jones & Bartlett Learning. ISBN 978-0-7637-8607-6.

- Berg, Jeremy M.; Tymoczko, John L.; Stryer, Lubert (2000), Biochemistry (5th ed.), New York: WH Freeman & Co., ISBN 0-7167-6766-X

- Voet, Donald; Voet, Judith; Pratt, Charlotte (2013). Fundamentals of Biochemistry: Life at the Molecular Level, Fourth Edition. John Wiley & Sons, Inc. p. 380. ISBN 1118129180.

- Rhodes, David. "Enzyme Kinetics - Single Substrate, Uncompetitive Inhibition, Lineweaver-Burk Plot". Purdue University. Retrieved 31 August 2013.

Enzyme Catalysis and Kinetics

When a chemical reaction involving enzymes undergoes an increase in its rate, the phenomenon is said to be an enzyme catalysis. This chapter studies the mechanism of action and the processes involved in enzyme catalysis by using enzyme kinetics. The reader gains a thorough understanding of the way enzymes act under varying conditions.

Enzyme Catalysis

Enzyme catalysis is the increase in the rate of a chemical reaction by the active site of a protein. The protein catalyst (enzyme) may be part of a multi-subunit complex, and/or may transiently or permanently associate with a Cofactor (e.g. adenosine triphosphate). Catalysis of biochemical reactions in the cell is vital due to the very low reaction rates of the uncatalysed reactions at room temperature and pressure. A key driver of protein evolution is the optimization of such catalytic activities via protein dynamics.

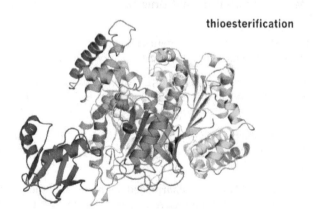

thioesterification

Visualization of ubiquitylation

The mechanism of enzyme catalysis is similar in principle to other types of chemical catalysis. By providing an alternative reaction route the enzyme reduces the energy required to reach the highest energy transition state of the reaction. The reduction of activation energy (Ea) increases the amount of reactant molecules that achieve a sufficient level of energy, such that they reach the activation energy and form the product. As with other catalysts, the enzyme is not consumed during the reaction (as a substrate is) but is recycled such that a single enzyme performs many rounds of catalysis.

Induced Fit

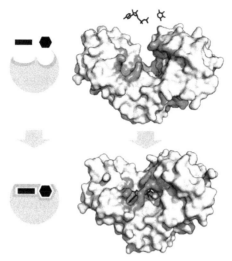

Enzyme changes shape by induced fit upon substrate binding to form enzyme-substrate complex. Hexokinase has a large induced fit motion that closes over the substrates adenosine triphosphate and xylose. Binding sites in blue, substrates in black and Mg^{2+} cofactor in yellow. (PDB: 2E2N, 2E2Q)

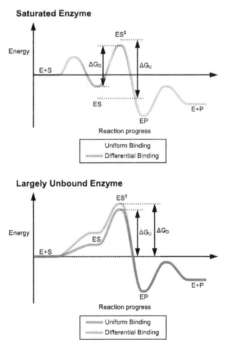

The different mechanisms of substrate binding

The favored model for the enzyme-substrate interaction is the induced fit model. This model proposes that the initial interaction between enzyme and substrate is relatively weak, but that these weak interactions rapidly induce conformational changes in the enzyme that strengthen binding.

The advantages of the induced fit mechanism arise due to the stabilizing effect of strong enzyme binding. There are two different mechanisms of substrate binding: uniform binding, which has strong substrate binding, and differential binding, which has strong transition state binding. The stabilizing effect of uniform binding increases both substrate and transition state binding affinity, while differential binding increases only transition state binding affinity. Both are used by enzymes and have been evolutionarily chosen to minimize the activation energy of the reaction. Enzymes that are saturated, that is, have a high affinity substrate binding, require differential binding to reduce the energy of activation, whereas small substrate unbound enzymes may use either differential or uniform binding.

These effects have led to most proteins using the differential binding mechanism to reduce the energy of activation, so most substrates have high affinity for the enzyme while in the transition state. Differential binding is carried out by the induced fit mechanism - the substrate first binds weakly, then the enzyme changes conformation increasing the affinity to the transition state and stabilizing it, so reducing the activation energy to reach it.

It is important to clarify, however, that the induced fit concept cannot be used to rationalize catalysis. That is, the chemical catalysis is defined as the reduction of $Ea^{\ddagger}$ (when the system is already in the $ES^{\ddagger}$) relative to $Ea^{\ddagger}$ in the uncatalyzed reaction in water (without the enzyme). The induced fit only suggests that the barrier is lower in the closed form of the enzyme but does not tell us what the reason for the barrier reduction is.

Induced fit may be beneficial to the fidelity of molecular recognition in the presence of competition and noise via the conformational proofreading mechanism . →→→editor

Mechanisms of An Alternative Reaction Route

These conformational changes also bring catalytic residues in the active site close to the chemical bonds in the substrate that will be altered in the reaction. After binding takes place, one or more mechanisms of catalysis lowers the energy of the reaction's transition state, by providing an alternative chemical pathway for the reaction. There are six possible mechanisms of "over the barrier" catalysis as well as a "through the barrier" mechanism:

Proximity and Orientation

This increases the rate of the reaction as enzyme-substrate interactions align reactive chemical groups and hold them close together. This reduces the entropy of the reactants and thus makes reactions such as ligations or addition reactions more favorable, there is a reduction in the overall loss of entropy when two reactants become a single product.

This effect is analogous to an effective increase in concentration of the reagents. The binding of the reagents to the enzyme gives the reaction intramolecular character, which gives a massive rate increase.

For example:

Similar reactions will occur far faster if the reaction is intramolecular.

Intermolecular

$k_1 = 4 \times 10^{-6}$ s-1M-1

Intramolecular

$k_2 = 0.8$ s-1

The effective concentration of acetate in the intramolecular reaction can be estimated as $k_2/k_1 = 2 \times 10^5$ Molar.

However, the situation might be more complex, since modern computational studies have established that traditional examples of proximity effects cannot be related directly to enzyme entropic effects. Also, the original entropic proposal has been found to largely overestimate the contribution of orientation entropy to catalysis.

Proton Donors or Acceptors

Proton donors and acceptors, i.e. acids and base may donate and accept protons in order to stabilize developing charges in the transition state.This typically has the effect of activating nucleophile and electrophile groups, or stabilizing leaving groups. Histidine is often the residue involved in these acid/base reactions, since it has a pKa close to neutral pH and can therefore both accept and donate protons.

Many reaction mechanisms involving acid/base catalysis assume a substantially altered pKa. This alteration of pKa is possible through the local environment of the residue.

Conditions	Acids	Bases
Hydrophobic environment	Increase pKa	Decrease pKa
Adjacent residues of like charge	Increase pKa	Decrease pKa
Salt bridge (and hydrogen bond) formation	Decrease pKa	Increase pKa

pKa can also be influenced significantly by the surrounding environment, to the extent that residues which are basic in solution may act as proton donors, and vice versa.

It is important to clarify that the modification of the pKa's is a pure part of the electrostatic mechanism. Furthermore, the catalytic effect of the above example is mainly associated with the reduction of the pKa of the oxyanion and the increase in the pKa of the histidine, while the proton transfer from the serine to the histidine is not catalyzed significantly, since it is not the rate determining barrier.

Electrostatic Catalysis

Stabilization of charged transition states can also be by residues in the active site forming ionic bonds (or partial ionic charge interactions) with the intermediate. These bonds can either come from acidic or basic side chains found on amino acids such as lysine, arginine, aspartic acid or glutamic acid or come from metal cofactors such as zinc. Metal ions are particularly effective and can reduce the pKa of water enough to make it an effective nucleophile.

Systematic computer simulation studies established that electrostatic effects give, by far, the largest contribution to catalysis. In particular, it has been found that enzyme provides an environment which is more polar than water, and that the ionic transition states are stabilized by fixed dipoles. This is very different from transition state stabilization in water, where the water molecules must pay with "reorganization energy". In order to stabilize ionic and charged states. Thus, the catalysis is associated with the fact that the enzyme polar groups are preorganized

The magnitude of the electrostatic field exerted by an enzyme's active site has been shown to be highly correlated with the enzyme's catalytic rate enhancement

Binding of substrate usually excludes water from the active site, thereby lowering the local dielectric constant to that of an organic solvent. This strengthens the electrostatic interactions between the charged/polar substrates and the active sites. In addition, studies have shown that the charge distributions about the active sites are arranged so as to stabilize the transition states of the catalyzed reactions. In several enzymes, these charge distributions apparently serve to guide polar substrates toward their binding sites so that the rates of these enzymatic reactions are greater than their apparent diffusion-controlled limits.

Covalent Catalysis

Covalent catalysis involves the substrate forming a transient covalent bond with residues in the enzyme active site or with a cofactor. This adds an additional covalent intermediate to the reaction, and helps to reduce the energy of later transition states of the reaction. The covalent bond must, at a later stage in the reaction, be broken to regenerate the enzyme. This mechanism is utilised by the catalytic triad of enzymes such as proteases like chymotrypsin and trypsin, where an acyl-enzyme intermediate is formed. An alternative mechanism is schiff base formation using the free amine from a lysine residue, as seen in the enzyme aldolase during glycolysis.

Some enzymes utilize non-amino acid cofactors such as pyridoxal phosphate (PLP) or thiamine pyrophosphate (TPP) to form covalent intermediates with reactant molecules. Such covalent intermediates function to reduce the energy of later transition states, similar to how covalent intermediates formed with active site amino acid residues allow stabilization, but the capabilities of cofactors allow enzymes to carryout reactions that amino acid side residues alone could not. Enzymes utilizing such cofactors include the PLP-dependent enzyme aspartate transaminase and the TPP-dependent enzyme pyruvate dehydrogenase.

Rather than lowering the activation energy for a reaction pathway, covalent catalysis provides an alternative pathway for the reaction (via to the covalent intermediate) and so is distinct from true catalysis. For example, the energetics of the covalent bond to the serine molecule in chymotrypsin should be compared to the well-understood covalent bond to the nucleophile in the uncatalyzed solution reaction. A true proposal of a covalent catalysis (where the barrier is lower than the corresponding barrier in solution) would require, for example, a partial covalent bond to the transition state by an enzyme group (e.g., a very strong hydrogen bond), and such effects do not contribute significantly to catalysis.

Metal Ion Catalysis

The presence of a metal ion in the active site participates in catalysis by coordinating charge stabilization and shielding. Because of a metal's positive charge, only negative charges can be stabilized through metal ions. Metal ions can also act to ionize water by acting as a Lewis acid. Metal ions may also be agents of oxidation and reduction.

Bond Strain

This is the principal effect of induced fit binding, where the affinity of the enzyme to the transition state is greater than to the substrate itself. This induces structural rearrangements which strain substrate bonds into a position closer to the conformation of the transition state, so lowering the energy difference between the substrate and transition state and helping catalyze the reaction.

However, the strain effect is, in fact, a ground state destabilization effect, rather than transition state stabilization effect. Furthermore, enzymes are very flexible and they cannot apply large strain effect.

In addition to bond strain in the substrate, bond strain may also be induced within the enzyme itself to activate residues in the active site.

Quantum Tunneling

These traditional "over the barrier" mechanisms have been challenged in some cases by models and observations of "through the barrier" mechanisms (quantum tunneling). Some enzymes operate with kinetics which are faster than what would be predicted by the classical $\Delta G^{\ddagger}$. In "through the barrier" models, a proton or an electron can tunnel through activation barriers. Quantum tunneling for protons has been observed in tryptamine oxidation by aromatic amine dehydrogenase.

Interestingly, quantum tunneling does not appear to provide a major catalytic advantage, since the tunneling contributions are similar in the catalyzed and the uncatalyzed reactions in solution. However, the tunneling contribution (typically enhancing rate constants by a factor of ~1000 compared to the rate of reaction for the classical 'over the barrier' route) is likely crucial to the viability of biological organisms. This emphasizes the general importance of tunneling reactions in biology.

In 1971-1972 the first quantum-mechanical model of enzyme catalysis was formulated.

Active Enzyme

The binding energy of the enzyme-substrate complex cannot be considered as an external energy which is necessary for the substrate activation. The enzyme of high energy content may firstly transfer some specific energetic group X_1 from catalytic site of the enzyme to the final place of the first bound reactant, then another group X_2 from the second bound reactant (or from the second group of the single reactant) must be transferred to active site to finish substrate conversion to product and enzyme regeneration.

We can present the whole enzymatic reaction as a two coupling reactions:

$$S_1 + EX_1 -> S_1 EX_1 -> P_1 + EP_2 \tag{1}$$

$$S_2 + EP_2 \rightarrow S_2 EP_2 \rightarrow P_2 + EX_2 \tag{2}$$

It may be seen from reaction (1) that the group X_1 of the active enzyme appears in the product due to possibility of the exchange reaction inside enzyme to avoid both electrostatic inhibition and repulsion of atoms. So we represent the active enzyme as a powerful reactant of the enzymatic reaction. The reaction (2) shows incomplete conversion of the substrate because its group X_2 remains inside enzyme. This approach as idea had formerly proposed relying on the hypothetical extremely high enzymatic conversions (catalytically perfect enzyme).

The crucial point for the verification of the present approach is that the catalyst must be a complex of the enzyme with the transfer group of the reaction. This chemical aspect is supported by the well-studied mechanisms of the several enzymatic reactions. Let us consider the reaction of peptide bond hydrolysis catalyzed by a pure protein α-chymotrypsin (an enzyme acting without a cofactor), which is a well-studied member of the serine proteases family.

We present the experimental results for this reaction as two chemical steps:

$$S_1 + EH -> P_1 + EP_2 \tag{3}$$

$$EP_2 + H - O - H -> EH + P_2 \tag{4}$$

where S_1 is a polypeptide, P_1 and P_2 are products. The first chemical step (3) includes the formation of a covalent acyl-enzyme intermediate. The second step (4) is the deacylation step. It is important to note that the group H+, initially found on the enzyme, but not in water, appears in the product before the step of hydrolysis, therefore it may be considered as an additional group of the enzymatic reaction.

Thus, the reaction (3) shows that the enzyme acts as a powerful reactant of the reaction. According to the proposed concept, the H transport from the enzyme promotes the first reactant conversion, breakdown of the first initial chemical bond (between groups P_1 and P_2). The step of hydrolysis leads to a breakdown of the second chemical bond and regeneration of the enzyme.

The proposed chemical mechanism does not depend on the concentration of the substrates or products in the medium. However, a shift in their concentration mainly causes free energy changes in the first and final steps of the reactions (1) and (2) due to the changes in the free energy content of every molecule, whether S or P, in water solution. This approach is in accordance with the following mechanism of muscle contraction. The final step of ATP hydrolysis in skeletal muscle is the product release caused by the association of myosin heads with actin. The closing of the actin-binding cleft during the association reaction is structurally coupled with the opening of the nucleotide-binding pocket on the myosin active site.

Notably, the final steps of ATP hydrolysis include the fast release of phosphate and the slow release of ADP. The release of a phosphate anion from bound ADP anion into water solution may be considered as an exergonic reaction because the phosphate anion has low molecular mass.

Thus, we arrive at the conclusion that the primary release of the inorganic phosphate $H_2PO_4^-$ leads to transformation of a significant part of the free energy of ATP hydrolysis into the kinetic energy of the solvated phosphate, producing active streaming. This assumption of a local mechano-chemical transduction is in accord with Tirosh's mechanism of muscle contraction, where the muscle force derives from an integrated action of active streaming created by ATP hydrolysis.

Examples of Catalytic Mechanisms

In reality, most enzyme mechanisms involve a combination of several different types of catalysis.

Triose Phosphate Isomerase

Triose phosphate isomerase (EC 5.3.1.1) catalyses the reversible interconvertion of the two triose phosphates isomers dihydroxyacetone phosphate and D-glyceraldehyde 3-phosphate.

Trypsin

Trypsin (EC 3.4.21.4) is a serine protease that cleaves protein substrates after lysine or arginine residues using a catalytic triad to perform covalent catalysis, and an oxyanion hole to stabilise charge-buildup on the transition states.

Aldolase

Aldolase (EC 4.1.2.13) catalyses the breakdown of fructose 1,6-bisphosphate (F-1,6-BP) into glyceraldehyde 3-phosphate and dihydroxyacetone phosphate (DHAP).

Enzyme Diffusivity

The advent of single-molecule studies led in the 2010s to the observation that the movement of untethered enzymes increases with increasing substrate concentration and increasing reaction enthalpy. Subsequent observations suggest that this increase in diffusivity is driven by transient displacement of the enzyme's center of mass, resulting in a "recoil effect that propels the enzyme".

Reaction Similarity

Similarity between enzymatic reactions (EC) can be calculated by using bond changes, reaction centres or substructure metrics (EC-BLAST).

Enzyme Kinetics

Enzyme kinetics is the study of the chemical reactions that are catalysed by enzymes. In enzyme kinetics, the reaction rate is measured and the effects of varying the conditions of the reaction are investigated. Studying an enzyme's kinetics in this way can reveal the catalytic mechanism of this enzyme, its role in metabolism, how its activity is controlled, and how a drug or an agonist might inhibit the enzyme.

Enzymes are usually protein molecules that manipulate other molecules — the enzymes' substrates. These target molecules bind to an enzyme's active site and are transformed into products through a series of steps known as the enzymatic mechanism

These mechanisms can be divided into single-substrate and multiple-substrate mechanisms. Kinetic studies on enzymes that only bind one substrate, such as triosephosphate isomerase, aim to measure the affinity with which the enzyme binds this substrate and the turnover rate. Some other examples of enzymes are phosphofructokinase and hexokinase, both of which are important for cellular respiration (glycolysis).

When enzymes bind multiple substrates, such as dihydrofolate reductase (shown right), enzyme kinetics can also show the sequence in which these substrates bind and the sequence in which products are released. An example of enzymes that bind a single substrate and release multiple products are proteases, which cleave one protein substrate into two polypeptide products. Others join two substrates together, such as DNA polymerase linking a nucleotide to DNA. Although these mechanisms are often a complex series of steps, there is typically one *rate-determining step* that determines the overall kinetics. This rate-determining step may be a chemical reaction or a conformational change of the enzyme or substrates, such as those involved in the release of product(s) from the enzyme.

Dihydrofolate reductase from *E. coli* with its two substrates dihydrofolate (right) and NADPH (left), bound in the active site. The protein is shown as a ribbon diagram, with alpha helices in red, beta sheets in yellow and loops in blue. Generated from 7DFR.

$$E + S \; \ulcorner \; ES \; \ulcorner \; ES^* \; \ulcorner \; EP \; \ulcorner \; E + P$$

Knowledge of the enzyme's structure is helpful in interpreting kinetic data. For example, the structure can suggest how substrates and products bind during catalysis; what changes occur during the reaction; and even the role of particular amino acid residues in the mechanism. Some enzymes change shape significantly during the mechanism; in such cases, it is helpful to determine the enzyme structure with and without bound substrate analogues that do not undergo the enzymatic reaction.

Not all biological catalysts are protein enzymes; RNA-based catalysts such as ribozymes and ribosomes are essential to many cellular functions, such as RNA splicing and translation. The main difference between ribozymes and enzymes is that RNA catalysts are composed of nucleotides, whereas enzymes are composed of amino acids. Ribozymes also perform a more limited set of reactions, although their reaction mechanisms and kinetics can be analysed and classified by the same methods.

General Principles

The reaction catalysed by an enzyme uses exactly the same reactants and produces exactly the same products as the uncatalysed reaction. Like other catalysts, enzymes do not alter the position of equilibrium between substrates and products. However, unlike uncatalysed chemical reactions, enzyme-catalysed reactions display saturation kinetics. For a given enzyme concentration and for relatively low substrate concentrations, the reaction rate increases linearly with substrate concentration; the enzyme molecules are largely free to catalyse the reaction, and increasing substrate concentration means an increasing rate at which the enzyme and substrate molecules encounter one another. However, at relatively high substrate concentrations, the reaction rate asymptotically approaches the theoretical maximum; the enzyme active sites are almost all occupied and the reaction rate is determined by the intrinsic turnover rate of the enzyme. The substrate concentration midway between these two limiting cases is denoted by K_M.

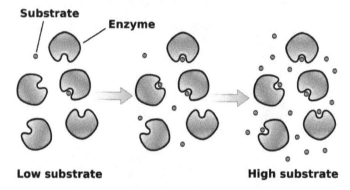

As larger amounts of substrate are added to a reaction, the available enzyme binding sites become filled to the limit of . Beyond this limit the enzyme is saturated with substrate and the reaction rate ceases to increase.

The two most important kinetic properties of an enzyme are how quickly the enzyme becomes saturated with a particular substrate, and the maximum rate it can achieve. Knowing these properties suggests what an enzyme might do in the cell and can show how the enzyme will respond to changes in these conditions.

Enzyme Assays

Enzyme assays are laboratory procedures that measure the rate of enzyme reactions. Because enzymes are not consumed by the reactions they catalyse, enzyme assays usually follow changes in the concentration of either substrates or products to measure the rate of reaction. There are many methods of measurement. Spectrophotometric assays observe change in the absorbance of light between products and reactants; radiometric assays involve the incorporation or release of radioactivity to measure the amount of product made over time. Spectrophotometric assays are most convenient since they allow the rate of the reaction to be measured continuously. Although radiometric assays require the removal and counting of samples (i.e., they are discontinuous assays) they are usually extremely sensitive and can measure very low levels of enzyme activity. An analogous approach is to use mass spectrometry to monitor the incorporation or release of stable isotopes as substrate is converted into product.

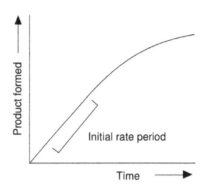

Progress curve for an enzyme reaction. The slope in the initial rate period is the initial rate of reaction v. The Michaelis–Menten equation describes how this slope varies with the concentration of substrate.

The most sensitive enzyme assays use lasers focused through a microscope to observe changes in single enzyme molecules as they catalyse their reactions. These measurements either use changes in the fluorescence of cofactors during an enzyme's reaction mechanism, or of fluorescent dyes added onto specific sites of the protein to report movements that occur during catalysis. These studies are providing a new view of the kinetics and dynamics of single enzymes, as opposed to traditional enzyme kinetics, which observes the average behaviour of populations of millions of enzyme molecules.

An example progress curve for an enzyme assay is shown above. The enzyme produces product at an initial rate that is approximately linear for a short period after the start of the reaction. As the reaction proceeds and substrate is consumed, the rate continuously slows (so long as substrate is not still at saturating levels). To measure the initial (and maximal) rate, enzyme assays are typically carried out while the reaction has progressed only a few percent towards total completion. The length of the initial rate period depends on the assay conditions and can range from milliseconds to hours. However, equipment for rapidly mixing liquids allows fast kinetic measurements on initial rates of less than one second. These very rapid assays are essential for measuring pre-steady-state kinetics, which are discussed below.

Most enzyme kinetics studies concentrate on this initial, approximately linear part of enzyme reactions. However, it is also possible to measure the complete reaction curve and fit this data to a non-linear rate equation. This way of measuring enzyme reactions is called progress-curve analysis. This approach is useful as an alternative to rapid kinetics when the initial rate is too fast to measure accurately.

Single-Substrate Reactions

Enzymes with single-substrate mechanisms include isomerases such as triosephos-phateisomerase or bisphosphoglycerate mutase, intramolecular lyases such as ade-nylate cyclase and the hammerhead ribozyme, an RNA lyase. However, some enzymes that only have a single substrate do not fall into this category of mechanisms. Catalase is an example of this, as the enzyme reacts with a first molecule of hydrogen peroxide substrate, becomes oxidised and is then reduced by a second molecule of substrate. Al-though a single substrate is involved, the existence of a modified enzyme intermediate means that the mechanism of catalase is actually a ping–pong mechanism, a type of mechanism that is discussed in the *Multi-substrate reactions* section below.

Michaelis–Menten Kinetics

As enzyme-catalysed reactions are saturable, their rate of catalysis does not show a lin-ear response to increasing substrate. If the initial rate of the reaction is measured over a range of substrate concentrations (denoted as [S]), the reaction rate (v) increases as [S] increases, as shown on the right. However, as [S] gets higher, the enzyme becomes saturated with substrate and the rate reaches V_{max}, the enzyme's maximum rate.

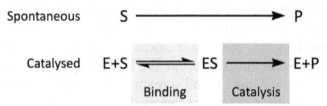

A chemical reaction mechanism with or without enzyme catalysis. The enzyme (E) binds substrate (S) to produce product (P).

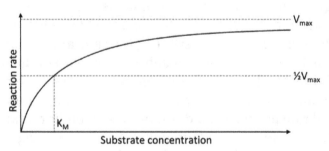

Saturation curve for an enzyme reaction showing the relation between the substrate concentration and reaction rate.

The Michaelis–Menten kinetic model of a single-substrate reaction is shown on the right. There is an initial bimolecular reaction between the enzyme E and substrate S to form the enzyme–substrate complex ES. The rate of enzymatic reaction increases with the increase of the substrate concentration up to a certain level called V_{max}; at V_{max}, in-crease in substrate concentration does not cause any increase in reaction rate as there no more enzyme (E) available for reacting with substrate (S). Here, the rate of reaction becomes dependent on the ES complex and the reaction becomes a unimolecular re-action with an order of zero. Though the enzymatic mechanism for the unimolecular reaction $ES \xrightarrow{k_{cat}} E + P$ can be quite complex, there is typically one rate-determining enzymatic step that allows this reaction to be modelled as a single catalytic step with an apparent unimolecular rate constant k_{cat}. If the reaction path proceeds over one or several intermediates, k_{cat} will be a function of several elementary rate constants, whereas in the simplest case of a single elementary reaction (e.g. no intermediates) it will be identical to the elementary unimolecular rate constant k_2. The apparent unimolecular rate constant k_{cat} is also called *turnover number* and denotes the maximum number of enzymatic reactions catalysed per second.

The Michaelis–Menten equation describes how the (initial) reaction rate v_0 depends on the position of the substrate-binding equilibrium and the rate constant k_2.

$$v_0 = \frac{V_{max}[S]}{K_M + [S]} \quad \text{(Michaelis–Menten equation)}$$

with the constants

$$K_M \overset{def}{=} \frac{k_2 + k_{-1}}{k_1} \approx K_D$$
$$V_{max} \overset{def}{=} k_{cat}[E]_{tot}$$

This Michaelis–Menten equation is the basis for most single-substrate enzyme kinetics. Two crucial assumptions underlie this equation (apart from the general assumption about the mechanism only involving no intermediate or product inhibition, and there is no allostericity or cooperativity). The first assumption is the so-called quasi-steady-state assumption (or pseudo-steady-state hypothesis), namely that the concentration of the substrate-bound enzyme (and hence also the unbound enzyme) changes much more slowly than those of the product and substrate and thus the change over time of the complex can be set to zero $d[ES]/dt = 0$. The second assumption is that the total enzyme concentration does not change over time, thus $[E]_{tot} = [E] + [ES] = const$. A complete derivation can be found here.

The Michaelis constant K_M is experimentally defined as the concentration at which the rate of the enzyme reaction is half V_{max}, which can be verified by substituting $[S] = K_m$ into the Michaelis–Menten equation and can also be seen graphically. If the rate-determining enzymatic step is slow compared to substrate dissociation ($k_2 \ll k_{-1}$), the Michaelis constant K_M is roughly the dissociation constant K_D of the ES complex.

If $[S]$ is small compared to K_M then the term $[S]/(K_M + [S]) \approx [S]/K_M$ and also very little ES complex is formed, thus $[E]_0 \approx [E]$. Therefore, the rate of product formation is

$$v_0 \approx \frac{k_{cat}}{K_M}[E][S] \qquad \text{if} [S] \ll K_M$$

Thus the product formation rate depends on the enzyme concentration as well as on the substrate concentration, the equation resembles a bimolecular reaction with a corresponding pseudo-second order rate constant k_2 / K_M. This constant is a measure of catalytic efficiency. The most efficient enzymes reach a k_2 / K_M in the range of $10^8 - 10^{10}$ M^{-1} s^{-1}. These enzymes are so efficient they effectively catalyse a reaction each time they encounter a substrate molecule and have thus reached an upper theoretical limit for efficiency (diffusion limit); and are sometimes referred to as kinetically perfect enzymes.

Direct Use of The Michaelis–Menten Equation for Time Course Kinetic Analysis

The observed velocities predicted by the Michaelis–Menten equation can be used to directly model the time course disappearance of substrate and the production of product through incorporation of the Michaelis–Menten equation into the equation for first order chemical kinetics. This can only be achieved however if one recognises the problem associated with the use of Euler's number in the description of first order chemical kinetics. i.e. e^{-k} is a split constant that introduces a systematic error into calculations and can be rewritten as a single constant which represents the remaining substrate after each time period.

$$[S] = [S]_0 (1-k)^t$$

$$[S] = [S]_0 (1-v/[S]_0)^t$$

$$[S] = [S]_0 (1-(V_{max}[S]_0 / (K_M + [S]_0)/[S]_0))^t$$

In 1983 Stuart Beal (and also independently Santiago Schnell and Claudio Mendoza in 1997) derived a closed form solution for the time course kinetics analysis of the Michaelis-Menten mechanism. The solution, known as the Schnell-Mendoza equation, has the form:

$$\frac{[S]}{K_M} = W[F(t)]$$

where W[] is the Lambert-W function. and where F(t) is

$$F(t) = \frac{[S]_0}{K_M} \exp\left(\frac{[S]_0}{K_M} - \frac{V_{max}}{K_M} t\right)$$

This equation is encompassed by the equation below, obtained by Berberan-Santos (MATCH Commun. Math. Comput. Chem. 63 (2010) 283), which is also valid when the initial substrate concentration is close to that of enzyme,

$$\frac{[S]}{K_M} = W[F(t)] - \frac{V_{max}}{k_{cat}K_M} \frac{W[F(t)]}{1+W[F(t)]}$$

where W[] is again the Lambert-W function.

Linear Plots of The Michaelis–Menten Equation

The plot of v versus [S] above is not linear; although initially linear at low [S], it bends over to saturate at high [S]. Before the modern era of nonlinear curve-fitting on computers, this nonlinearity could make it difficult to estimate K_M and V_{max} accurately. Therefore, several researchers developed linearisations of the Michaelis–Menten equation, such as the Lineweaver–Burk plot, the Eadie–Hofstee diagram and the Hanes–Woolf plot. All of these linear representations can be useful for visualising data, but none should be used to determine kinetic parameters, as computer software is readily available that allows for more accurate determination by nonlinear regression methods.

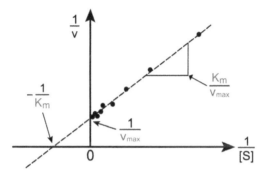

Lineweaver–Burk or double-reciprocal plot of kinetic data, showing the significance of the axis intercepts and gradient.

The Lineweaver–Burk plot or double reciprocal plot is a common way of illustrating kinetic data. This is produced by taking the reciprocal of both sides of the Michaelis–Menten equation. As shown on the right, this is a linear form of the Michaelis–Menten equation and produces a straight line with the equation $y = mx + c$ with a y-intercept equivalent to $1/V_{max}$ and an x-intercept of the graph representing $-1/K_M$.

$$\frac{1}{v} = \frac{K_M}{V_{max}[S]} + \frac{1}{V_{max}}$$

Naturally, no experimental values can be taken at negative $1/[S]$; the lower limiting value $1/[S] = 0$ (the y-intercept) corresponds to an infinite substrate concentration, where $1/v = 1/V_{max}$ as shown at the right; thus, the x-intercept is an extrapolation of the experimental data taken at positive concentrations. More generally, the Lineweaver–Burk plot skews the importance of measurements taken at low substrate concentrations and, thus, can yield inaccurate estimates of V_{max} and K_M. A more accurate linear plotting method is the Eadie-Hofstee plot. In this case, v is plotted against $v/[S]$. In the third common linear representation, the Hanes-Woolf plot, $[S]/v$ is plotted against [S]. In general, data normalisation can help diminish the amount of experimental work and can increase the reliability of the output, and is suitable for both graphical and numerical analysis.

Practical Significance of Kinetic Constants

The study of enzyme kinetics is important for two basic reasons. Firstly, it helps explain how enzymes work, and secondly, it helps predict how enzymes behave in living organisms. The kinetic constants defined above, K_M and V_{max}, are critical to attempts to understand how enzymes work together to control metabolism.

Making these predictions is not trivial, even for simple systems. For example, oxaloacetate is formed by malate dehydrogenase within the mitochondrion. Oxaloacetate can then be consumed by citrate synthase, phosphoenolpyruvate carboxykinase or aspartate aminotransferase, feeding into the citric acid cycle, gluconeogenesis or aspartic acid biosynthesis, respectively. Being able to predict how much oxaloacetate goes into which pathway requires knowledge of the concentration of oxaloacetate as well as the concentration and kinetics of each of these enzymes. This aim of predicting the behaviour of metabolic pathways reaches its most complex expression in the synthesis of huge amounts of kinetic and gene expression data into mathematical models of entire organisms. Alternatively, one useful simplification of the metabolic modelling problem is to ignore the underlying enzyme kinetics and only rely on information about the reaction network's stoichiometry, a technique called flux balance analysis.

Michaelis–Menten Kinetics with Intermediate

One could also consider the less simple case

$$E + S \underset{k_{-1}}{\overset{k_1}{\rightleftharpoons}} ES \xrightarrow{k_2} EI \xrightarrow{k_3} E + P$$

where a complex with the enzyme and an intermediate exists and the intermediate is converted into product in a second step. In this case we have a very similar equation

$$v_0 = k_{cat} \frac{[S][E]_0}{K'_M + [S]}$$

but the constants are different

$$K'_M \overset{def}{=} \frac{k_3}{k_2 + k_3} K_M = \frac{k_3}{k_2 + k_3} \cdot \frac{k_2 + k_{-1}}{k_1}$$

$$k_{cat} \overset{def}{=} \frac{k_3 k_2}{k_2 + k_3}$$

We see that for the limiting case $k_3 \gg k_2$, thus when the last step from $EI \to E + P$ is much faster than the previous step, we get again the original equation. Mathematically we have then $K'_M \approx K_M$ and $k_{cat} \approx k_2$.

Multi-Substrate Reactions

Multi-substrate reactions follow complex rate equations that describe how the substrates bind and in what sequence. The analysis of these reactions is much simpler if the concentration of substrate A is kept constant and substrate B varied. Under these conditions, the enzyme behaves just like a single-substrate enzyme and a plot of v by [S] gives apparent K_M and V_{max} constants for substrate B. If a set of these measurements is performed at different fixed concentrations of A, these data can be used to work out what the mechanism of the reaction is. For an enzyme that takes two substrates A and B and turns them into two products P and Q, there are two types of mechanism: ternary complex and ping–pong.

Ternary-Complex Mechanisms

In these enzymes, both substrates bind to the enzyme at the same time to produce an EAB ternary complex. The order of binding can either be random (in a random mechanism) or substrates have to bind in a particular sequence (in an ordered mechanism). When a set of v by [S] curves (fixed A, varying B) from an enzyme with a ternary-complex mechanism are plotted in a Lineweaver–Burk plot, the set of lines produced will intersect.

Random-order ternary-complex mechanism for an enzyme reaction. The reaction path is shown as a line and enzyme intermediates containing substrates A and B or products P and Q are written below the line.

Enzymes with ternary-complex mechanisms include glutathione S-transferase, dihydrofolate reductase and DNA polymerase. The following links show short animations of the ternary-complex mechanisms of the enzymes dihydrofolate reductase[β] and DNA polymerase[γ].

Ping–Pong Mechanisms

As shown on the right, enzymes with a ping-pong mechanism can exist in two states, E and a chemically modified form of the enzyme E*; this modified enzyme is known as an intermediate. In such mechanisms, substrate A binds, changes the enzyme to E* by, for example, transferring a chemical group to the active site, and is then released. Only after the first substrate is released can substrate B bind and react with the modified enzyme, regenerating the unmodified E form. When a set of v by [S] curves (fixed A, varying B) from an enzyme with a ping–pong mechanism are plotted in a Lineweaver–Burk plot, a set of parallel lines will be produced. This is called a secondary plot.

Ping–pong mechanism for an enzyme reaction. Intermediates contain substrates A and B or products P and Q.

Enzymes with ping–pong mechanisms include some oxidoreductases such as thiore-doxin peroxidase, transferases such as acylneuraminate cytidylyltransferase and serine proteases such as trypsin and chymotrypsin. Serine proteases are a very common and diverse family of enzymes, including digestive enzymes (trypsin, chymotrypsin, and elastase), several enzymes of the blood clotting cascade and many others. In these ser-ine proteases, the E* intermediate is an acyl-enzyme species formed by the attack of an active site serine residue on a peptide bond in a protein substrate. A short animation showing the mechanism of chymotrypsin is linked here.[6]

Non-Michaelis–Menten Kinetics

Some enzymes produce a sigmoid v by [S] plot, which often indicates cooperative bind-ing of substrate to the active site. This means that the binding of one substrate molecule affects the binding of subsequent substrate molecules. This behavior is most common in multimeric enzymes with several interacting active sites. Here, the mechanism of cooperation is similar to that of hemoglobin, with binding of substrate to one active site altering the affinity of the other active sites for substrate molecules. Positive coopera-tivity occurs when binding of the first substrate molecule *increases* the affinity of the other active sites for substrate. Negative cooperativity occurs when binding of the first substrate *decreases* the affinity of the enzyme for other substrate molecules.

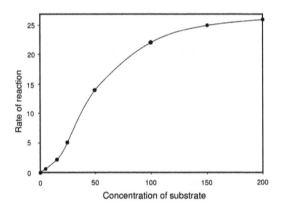

Saturation curve for an enzyme reaction showing sigmoid kinetics.

Allosteric enzymes include mammalian tyrosyl tRNA-synthetase, which shows neg-ative cooperativity, and bacterial aspartate transcarbamoylase and phosphofructoki-nase, which show positive cooperativity.

Cooperativity is surprisingly common and can help regulate the responses of enzymes to changes in the concentrations of their substrates. Positive cooperativity makes en-zymes much more sensitive to [S] and their activities can show large changes over a narrow range of substrate concentration. Conversely, negative cooperativity makes en-zymes insensitive to small changes in [S].

The Hill equation (biochemistry) is often used to describe the degree of cooperativity quantitatively in non-Michaelis–Menten kinetics. The derived Hill coefficient n measures how much the binding of substrate to one active site affects the binding of substrate to the other active sites. A Hill coefficient of <1 indicates negative cooperativity and a coefficient of >1 indicates positive cooperativity.

Pre-Steady-State Kinetics

In the first moment after an enzyme is mixed with substrate, no product has been formed and no intermediates exist. The study of the next few milliseconds of the reaction is called Pre-steady-state kinetics also referred to as Burst kinetics. Pre-steady-state kinetics is therefore concerned with the formation and consumption of enzyme–substrate intermediates (such as ES or E*) until their steady-state concentrations are reached.

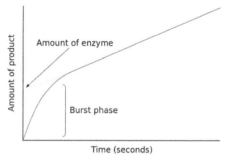

Pre-steady state progress curve, showing the burst phase of an enzyme reaction.

This approach was first applied to the hydrolysis reaction catalysed by chymotrypsin. Often, the detection of an intermediate is a vital piece of evidence in investigations of what mechanism an enzyme follows. For example, in the ping–pong mechanisms that are shown above, rapid kinetic measurements can follow the release of product P and measure the formation of the modified enzyme intermediate E*. In the case of chymotrypsin, this intermediate is formed by an attack on the substrate by the nucleophilic serine in the active site and the formation of the acyl-enzyme intermediate.

In the figure to the right, the enzyme produces E* rapidly in the first few seconds of the reaction. The rate then slows as steady state is reached. This rapid burst phase of the reaction measures a single turnover of the enzyme. Consequently, the amount of product released in this burst, shown as the intercept on the y-axis of the graph, also gives the amount of functional enzyme which is present in the assay.

Chemical mechanism

An important goal of measuring enzyme kinetics is to determine the chemical mechanism of an enzyme reaction, i.e., the sequence of chemical steps that transform substrate into product. The kinetic approaches discussed above will show at what rates intermediates are formed and inter-converted, but they cannot identify exactly what these intermediates are.

Kinetic measurements taken under various solution conditions or on slightly modified enzymes or substrates often shed light on this chemical mechanism, as they reveal the rate-determining step or intermediates in the reaction. For example, the breaking of a covalent bond to a hydrogen atom is a common rate-determining step. Which of the possible hydrogen transfers is rate determining can be shown by measuring the kinetic effects of substituting each hydrogen by deuterium, its stable isotope. The rate will change when the critical hydrogen is replaced, due to a primary kinetic isotope effect, which occurs because bonds to deuterium are harder to break than bonds to hydrogen. It is also possible to measure similar effects with other isotope substitutions, such as $^{13}C/^{12}C$ and $^{18}O/^{16}O$, but these effects are more subtle.

Isotopes can also be used to reveal the fate of various parts of the substrate molecules in the final products. For example, it is sometimes difficult to discern the origin of an oxygen atom in the final product; since it may have come from water or from part of the substrate. This may be determined by systematically substituting oxygen's stable isotope ^{18}O into the various molecules that participate in the reaction and checking for the isotope in the product. The chemical mechanism can also be elucidated by examining the kinetics and isotope effects under different pH conditions, by altering the metal ions or other bound cofactors, by site-directed mutagenesis of conserved amino acid residues, or by studying the behaviour of the enzyme in the presence of analogues of the substrate(s).

Enzyme Inhibition and Activation

Enzyme inhibitors are molecules that reduce or abolish enzyme activity, while enzyme activators are molecules that increase the catalytic rate of enzymes. These interactions can be either *reversible* (i.e., removal of the inhibitor restores enzyme activity) or *irreversible* (i.e., the inhibitor permanently inactivates the enzyme).

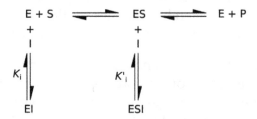

Kinetic scheme for reversible enzyme inhibitors.

Reversible Inhibitors

Traditionally reversible enzyme inhibitors have been classified as competitive, uncompetitive, or non-competitive, according to their effects on K_m and V_{max}. These different effects result from the inhibitor binding to the enzyme E, to the enzyme–substrate complex ES, or to both, respectively. The division of these classes arises from a problem in their derivation and results in the need to use two different binding constants for one binding event. The binding of an inhibitor and its effect on the enzymatic activity are two distinctly different things, another problem the traditional equations fail to acknowledge. In noncompetitive inhibition the binding of the inhibitor results in 100% inhibition of the enzyme only, and fails to consider the possibility of anything in between. The common form of the inhibitory term also obscures the relationship between the inhibitor binding to the enzyme and its relationship to any other binding term be it the Michaelis–Menten equation or a dose response curve associated with ligand receptor binding. To demonstrate the relationship the following rearrangement can be made:

$$\frac{V_{max}}{1+\dfrac{[I]}{K_i}}=\frac{V_{max}}{\dfrac{[I]+K_i}{K_i}}$$

Adding zero to the bottom ([I]-[I])

$$\frac{\dfrac{V_{max}}{[I]+K_i}}{[I]+K_i-[I]}$$

Dividing by [I]+K$_i$

$$\frac{\dfrac{V_{max}}{1}}{1-\dfrac{[I]}{[I]+K_i}}=V_{max}-V_{max}\frac{[I]}{[I]+K_i}$$

This notation demonstrates that similar to the Michaelis–Menten equation, where the rate of reaction depends on the percent of the enzyme population interacting with substrate

fraction of the enzyme population bound by substrate

$$\frac{[S]}{[S]+K_m}$$

fraction of the enzyme population bound by inhibitor

$$\frac{[I]}{[I]+K_i}$$

the effect of the inhibitor is a result of the percent of the enzyme population interacting with inhibitor. The only problem with this equation in its present form is that it assumes absolute inhibition of the enzyme with inhibitor binding, when in fact there can be a wide range of effects anywhere from 100% inhibition of substrate turn over to just >0%. To account for this the equation can be easily modified to allow for different degrees of inhibition by including a delta V_{max} term.

$$V_{max}-\Delta V_{max}\frac{[I]}{[I]+K_i}$$

Or

$$V_{max1}-(V_{max1}-V_{max2})\frac{[I]}{[I]+K_i}$$

This term can then define the residual enzymatic activity present when the inhibitor is interacting with individual enzymes in the population. However the inclusion of this term has the added value of allowing for the possibility of activation if the secondary V_{max} term turns out to be higher than the initial term. To account for the possibly of activation as well the notation can then be rewritten replacing the inhibitor "I" with a modifier term denoted here as "X".

$$V_{max1} - (V_{max1} - V_{max2})\frac{[X]}{[X]+K_x}$$

While this terminology results in a simplified way of dealing with kinetic effects relating to the maximum velocity of the Michaelis–Menten equation, it highlights potential problems with the term used to describe effects relating to the K_m. The K_m relating to the affinity of the enzyme for the substrate should in most cases relate to potential changes in the binding site of the enzyme which would directly result from enzyme inhibitor interactions. As such a term similar to the one proposed above to modulate V_{max} should be appropriate in most situations:

$$K_{m1} - (K_{m1} - K_{m2})\frac{[X]}{[X]+K_x}$$

Irreversible Inhibitors

Enzyme inhibitors can also irreversibly inactivate enzymes, usually by covalently modifying active site residues. These reactions, which may be called suicide substrates, follow exponential decay functions and are usually saturable. Below saturation, they follow first order kinetics with respect to inhibitor.

Mechanisms of Catalysis

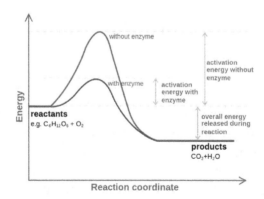

The energy variation as a function of reaction coordinate shows the stabilisation of the transition state by an enzyme.

The favoured model for the enzyme–substrate interaction is the induced fit model. This model proposes that the initial interaction between enzyme and substrate is relatively weak, but that these weak interactions rapidly induce conformational changes in the enzyme that strengthen binding. These conformational changes also bring catalytic residues in the active site close to the chemical bonds in the substrate that will be altered in the reaction. Conformational changes can be measured using circular dichroism or dual polarisation interferometry. After binding takes place, one or more mechanisms of catalysis lower the energy of the reaction's transition state by providing an alternative chemical pathway for the reaction. Mechanisms of catalysis include catalysis by bond strain; by proximity and orientation; by active-site proton donors or acceptors; covalent catalysis and quantum tunnelling.

Enzyme kinetics cannot prove which modes of catalysis are used by an enzyme. However, some kinetic data can suggest possibilities to be examined by other techniques. For example, a ping–pong mechanism with burst-phase pre-steady-state kinetics would suggest covalent catalysis might be important in this enzyme's mechanism. Alternatively, the observation of a strong pH effect on V_{max} but not K_m might indicate that a residue in the active site needs to be in a particular ionisation state for catalysis to occur.

History

In 1902 Victor Henri proposed a quantitative theory of enzyme kinetics, but at the time the experimental significance of the hydrogen ion concentration was not yet recognized. After Peter Lauritz Sørensen had defined the logarithmic pH-scale and introduced the concept of buffering in 1909 the German chemist Leonor Michaelis and his Canadian postdoc Maud Leonora Menten repeated Henri's experiments and confirmed his equation, which is now generally referred to as Michaelis-Menten kinetics (sometimes also *Henri-Michaelis-Menten kinetics*). Their work was further developed by G. E. Briggs and J. B. S. Haldane, who derived kinetic equations that are still widely considered today a starting point in modeling enzymatic activity.

The major contribution of the Henri-Michaelis-Menten approach was to think of enzyme reactions in two stages. In the first, the substrate binds reversibly to the enzyme, forming the enzyme-substrate complex. This is sometimes called the Michaelis complex. The enzyme then catalyzes the chemical step in the reaction and releases the product. The kinetics of many enzymes is adequately described by the simple Michaelis-Menten model, but all enzymes have internal motions that are not accounted for in the model and can have significant contributions to the overall reaction kinetics. This can be modeled by introducing several Michaelis-Menten pathways that are connected with fluctuating rates, which is a mathematical extension of the basic Michaelis Menten mechanism.

Software

ENZO

ENZO (Enzyme Kinetics) is a graphical interface tool for building kinetic models of enzyme catalyzed reactions. ENZO automatically generates the corresponding differential equations from a stipulated enzyme reaction scheme. These differential equations are processed by a numerical solver and a regression algorithm which fits the coefficients of differential equations to experimentally observed time course curves. ENZO allows rapid evaluation of rival reaction schemes and can be used for routine tests in enzyme kinetics.

References

- Voet, Donald; Judith Voet (2004). Biochemistry. John Wiley & Sons Inc. pp. 986–989. ISBN 0-471-25090-2.

- Walsh, Ryan (2012). "Ch. 17. Alternative Perspectives of Enzyme Kinetic Modeling". In Ekinci, Deniz. Medicinal Chemistry and Drug Design (PDF). InTech. pp. 357–371. ISBN 978-953-51-0513-8.

5

Metabolic Pathway of Enzyme

This chapter focuses on the metabolic pathways of oxidative phosphorylation, citric acid cycle, glycolysis, pentose phosphate pathway, fatty acid synthesis etc. Understanding the metabolic pathways helps grasp how an enzyme functions, the series of reactions that take place by the action of an enzyme and the facilitators of enzyme reactions. The topics discussed in the chapter are of great importance to broaden the existing knowledge on enzymes.

Metabolic Pathway

In biochemistry, a metabolic pathway is a linked series of chemical reactions occurring within a cell. The reactants, products, and intermediates of an enzymatic reaction are known as metabolites, which are modified by a sequence of chemical reactions catalyzed by enzymes. In a metabolic pathway, the product of one enzyme acts as the substrate for the next. These enzymes often require dietary minerals, vitamins, and other cofactors to function.

Different metabolic pathways function based on the position within a eukaryotic cell and the significance of the pathway in the given compartment of the cell. For instance, the citric acid cycle, electron transport chain, and oxidative phosphorylation all take place in the mitochondrial membrane. In contrast, glycolysis, pentose phosphate pathway, and fatty acid biosynthesis all occur in the cytosol of a cell.

There are two types of metabolic pathways that are characterized by their ability to either synthesize molecules with the utilization of energy (anabolic pathway) or break down of complex molecules by releasing energy in the process (catabolic pathway). The two pathways compliment each other in that the energy released from one is used up by the other. The degradative process of a catabolic pathway provides the energy required to conduct a biosynthesis of an anabolic pathway. In addition to the two distinct metabolic pathways is the amphibolic pathway, which can be either catabolic or anabolic based on the need for or the availability of energy.

Pathways are required for the maintenance of homeostasis within an organism and the flux of metabolites through a pathway is regulated depending on the needs of the cell and the availability of the substrate. The end product of a pathway may be used immediately, initiate another metabolic pathway or be stored for later use. The metabolism of a cell consists of an elaborate network of interconnected pathways that enable the synthesis and breakdown of molecules (anabolism and catabolism)

Overview

Net reactions of common metabolic pathways

Each metabolic pathway consists of a series of biochemical reactions that are connected by their intermediates: the products of one reaction are the substrates for subsequent reactions, and so on. Metabolic pathways are often considered to flow in one direction. Although all chemical reactions are technically reversible, conditions in the cell are often such that it is thermodynamically more favorable for flux to flow in one direction of a reaction. For example, one pathway may be responsible for the synthesis of a particular amino acid, but the breakdown of that amino acid may occur via a separate and distinct pathway. One example of an exception to this "rule" is the metabolism of glucose. Glycolysis results in the breakdown of glucose, but several reactions in the glycolysis pathway are reversible and participate in the re-synthesis of glucose (gluconeogenesis).

- Glycolysis was the first metabolic pathway discovered:

1. As glucose enters a cell, it is immediately phosphorylated by ATP to glucose 6-phosphate in the irreversible first step.

2. In times of excess lipid or protein energy sources, certain reactions in the glycolysis pathway may run in reverse in order to produce glucose 6-phosphate which is then used for storage as glycogen or starch.

- Metabolic pathways are often regulated by feedback inhibition.

- Some metabolic pathways flow in a 'cycle' wherein each component of the cycle is a substrate for the subsequent reaction in the cycle, such as in the Krebs Cycle.

- Anabolic and catabolic pathways in eukaryotes often occur independently of each other, separated either physically by compartmentalization within organelles or separated biochemically by the requirement of different enzymes and co-factors.

Catabolic Pathway (Catabolism)

A catabolic pathway is a series of reactions that bring about a net release of energy in the form of a high energy phosphate bond formed with the energy carriers Adenosine Diphosphate (ADP) and Guanosine Diphosphate (GDP) to produce Adenosine Triphosphate (ATP) and Guanosine Triphosphate (GTP), respectively. The net reaction is, therefore, thermodynamically favorable, for it results in a lower free energy for the final products. A catabolic pathway is an exergonic system that produces chemical energy in the form of ATP, GTP, NADH, NADPH, FADH2, etc. from energy containing sources such as carbohydrates, fats, and proteins. The end products are often carbon dioxide, water, and ammonia. Coupled with an endergonic reaction of anabolism, the cell can synthesize new macromolecules using the original precursors of the anabolic pathway. An example of a coupled reaction is the phosphorylation of fructose-6-phosphate to form the intermediate fructose-1,6-bisphosphate by the enzyme phsophofructokinase accompanied by the hydrolysis of ATP in the pathway of glycolysis. The resulting chemical reaction within the metabolic pathway is highly thermodynamically favorable and, as a result, irreversible in the cell.

$$Fructose-6-Phosphate+ATPFructose-1,6-Bisphosphate+ADP$$

Cellular Respiration

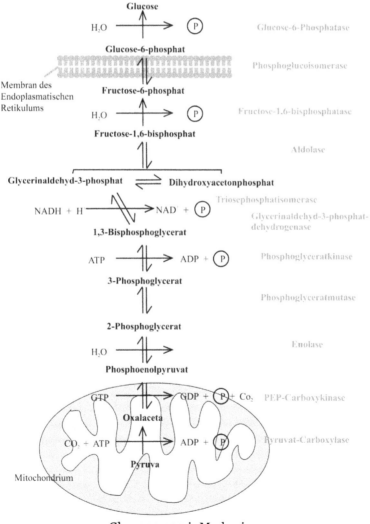

Gluconeogenesis Mechanism

A core set of energy-producing catabolic pathways occur within all living organisms in some form. These pathways transfer the energy released by breakdown of nutrients into ATP and other small molecules used for energy (e.g. GTP, NADPH, FADH). All cells can perform anaerobic respiration by glycolysis. Additionally, most organisms can perform more efficient aerobic respiration through the citric acid cycle and oxidative phosphorylation. Additionally plants, algae and cyanobacteria are able to use sunlight to anabolically synthesize compounds from non-living matter by photosynthesis.

Anabolic Pathway (Anabolism)

In contrast to the catabolic pathways, are the anabolic pathways that require an input of energy in order to conduct the construction of macromolecules such as polypeptides, nucleic acids, proteins, polysaccharides, and lipids. The isolated reaction of anabolism is unfavorable in a cell due to a positive Gibbs Free Energy $(+\Delta G)$; thus, an input of chemical energy through a coupling with an exergonic reaction is necessary. The coupled reaction of the catabolic pathway affects the thermodynamics of the reaction by lowering the overall activation energy of an anabolic pathway and allowing the reaction to take place. Otherwise, an endergonic reaction is non-spontaneous.

An anabolic pathway is a biosynthetic pathway, meaning that it combines smaller molecules to form larger and more complex ones. An example is the reversed pathway of glycolysis, otherwise known as gluconeogenesis, which occurs in the liver and sometimes in the kidney in order to maintain proper glucose concentration in the blood and to be able to supply the brain and muscle tissues with adequate amount of glucose. Although gluconeogenesis is similar to the reverse pathway of glycolysis, it contains three distinct enzymes from glycolysis that allow the pathway to occur spontaneously. An example of the pathway for gluconeogenesis is illustrated in the image titled "Gluconeogenesis Mechanism".

Amphibolic Pathway

An amphibolic pathway is one that can be either catabolic or anabolic based on the availability of or the need for energy. The currency of energy in a biological cell is adenosine triphosphate (ATP), which stores its energy in the phosphoanhydride bonds. The energy is utilized to conduct biosynthesis, facilitate movement, and regulate active transport inside of the cell. Examples of amphibolic pathways are the citric acid cycle and the glyoxylate cycle. These sets of chemical reactions contain both energy producing and utilizing pathways. To the right is an illustration of the amphibolic properties of the TCA cycle.

The glyoxylate shunt pathway is an alternative to the tricarboxylic acid (TCA) cycle, for it redirects the pathway of TCA to prevent full oxidation of carbon compounds, and to preserve high energy carbon sources as future energy sources. This pathway occurs only in plants and bacteria and transpires in the absence of glucose molecules.

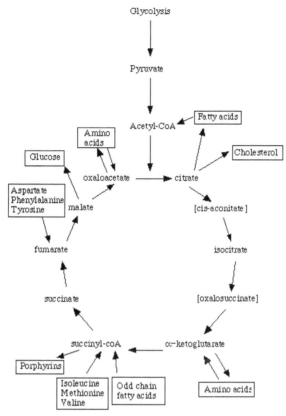

Amphibolic Properties of the Citric Acid Cycle

Regulation

The flux of the entire pathway is regulated by the rate-determining steps. These are the slowest steps in a network of reactions. The rate-limiting step occurs near the beginning of the pathway and is regulated by feedback inhibition, which ultimately controls the overall rate of the pathway. The metabolic pathway in the cell is regulated by covalent or non-covalent modifications. A covalent modification involves an addition or removal of a chemical bond, whereas a non-covalent modification (also known as allosteric regulation) is the binding of the regulator to the enzyme via hydrogen bonds, electrostatic interactions, and Van Der Waals forces.

The rate of turnover in a metabolic pathway, otherwise known as the metabolic flux, is regulated based on the stoichiometric reaction model, the utilization rate of metabolites, and the translocation pace of molecules across the lipid bilayer. The regulation methods are based on experiments involving 13C-labeling, which is then analyzed by Nuclear Magnetic Resonance (NMR) or gas chromatography-mass spectrometry (GC-MS)-derived mass compositions. The aforementioned techniques synthesize a statistical interpretation of mass distribution in proteinogenic amino acids to the catalytic activities of enzymes in a cell.

Citric Acid Cycle

The citric acid cycle – also known as the tricarboxylic acid (TCA) cycle or the Krebs cycle – is a series of chemical reactions used by all aerobic organisms to generate energy through the oxidation of acetyl-CoA derived from carbohydrates, fats and proteins into carbon dioxide and chemical energy in the form of adenosine triphosphate. In addition, the cycle provides precursors of certain amino acids as well as the reducing agent NADH that is used in numerous other biochemical reactions. Its central importance to many biochemical pathways suggests that it was one of the earliest established components of cellular metabolism and may have originated abiogenically.

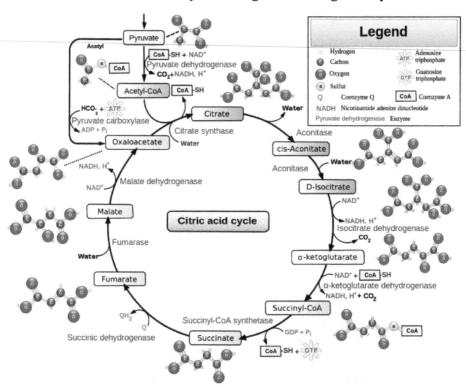

Overview of the citric acid cycle *(click to enlarge)*

The name of this metabolic pathway is derived from citric acid (a type of tricarboxylic acid) that is consumed and then regenerated by this sequence of reactions to complete the cycle. In addition, the cycle consumes acetate (in the form of acetyl-CoA) and water, reduces NAD^+ to NADH, and produces carbon dioxide as a waste byproduct. The NADH generated by the TCA cycle is fed into the oxidative phosphorylation (electron transport) pathway. The net result of these two closely linked pathways is the oxidation of nutrients to produce usable chemical energy in the form of ATP.

In eukaryotic cells, the citric acid cycle occurs in the matrix of the mitochondrion. In prokaryotic cells, such as bacteria which lack mitochondria, the TCA reaction sequence is performed in the cytosol with the proton gradient for ATP production being across the cell's surface (plasma membrane) rather than the inner membrane of the mitochondrion.

Discovery

Several of the components and reactions of the citric acid cycle were established in the 1930s by the research of the Nobel laureate Albert Szent-Györgyi, for which he received the Nobel Prize in 1937 for his discoveries pertaining to fumaric acid, a key component of the cycle. The citric acid cycle itself was finally identified in 1937 by Hans Adolf Krebs while at the University of Sheffield, for which he received the Nobel Prize for Physiology or Medicine in 1953.

Evolution

Components of the TCA cycle were derived from anaerobic bacteria, and the TCA cycle itself may have evolved more than once. Theoretically there are several alternatives to the TCA cycle; however, the TCA cycle appears to be the most efficient. If several TCA alternatives had evolved independently, they all appear to have converged to the TCA cycle.

Overview

The citric acid cycle is a key metabolic pathway that unifies carbohydrate, fat, and protein metabolism. The reactions of the cycle are carried out by 8 enzymes that completely oxidize acetate, in the form of acetyl-CoA, into two molecules each of carbon dioxide and water. Through catabolism of sugars, fats, and proteins, a two-carbon organic product acetate in the form of acetyl-CoA is produced which enters the citric acid cycle. The reactions of the cycle also convert three equivalents of nicotinamide adenine dinucleotide (NAD^+) into three equivalents of reduced NAD^+ (NADH), one equivalent of flavin adenine dinucleotide (FAD) into one equivalent of $FADH_2$, and one equivalent each of guanosine diphosphate (GDP) and inorganic phosphate (P_i) into one equivalent of guanosine triphosphate (GTP). The NADH and $FADH_2$ generated by the citric acid cycle are in turn used by the oxidative phosphorylation pathway to generate energy-rich adenosine triphosphate (ATP).

Structural diagram of acetyl-CoA. The portion in blue, on the left, is the acetyl group; the portion in black is coenzyme A.

One of the primary sources of acetyl-CoA is from the breakdown of sugars by glycolysis which yield pyruvate that in turn is decarboxylated by the enzyme pyruvate dehydrogenase generating acetyl-CoA according to the following reaction scheme:

- $CH_3C(=O)C(=O)O^-$ (pyruvate) + HSCoA + NAD^+ → $CH_3C(=O)SCoA$ (acetyl-CoA) + NADH + CO_2

The product of this reaction, acetyl-CoA, is the starting point for the citric acid cycle. Acetyl-CoA may also be obtained from the oxidation of fatty acids. Below is a schematic outline of the cycle:

- The citric acid cycle begins with the transfer of a two-carbon acetyl group from acetyl-CoA to the four-carbon acceptor compound (oxaloacetate) to form a six-carbon compound (citrate).

- The citrate then goes through a series of chemical transformations, losing two carboxyl groups as CO_2. The carbons lost as CO_2 originate from what was oxaloacetate, not directly from acetyl-CoA. The carbons donated by acetyl-CoA become part of the oxaloacetate carbon backbone after the first turn of the citric acid cycle. Loss of the acetyl-CoA-donated carbons as CO_2 requires several turns of the citric acid cycle. However, because of the role of the citric acid cycle in anabolism, they might not be lost, since many TCA cycle intermediates are also used as precursors for the biosynthesis of other molecules.

- Most of the energy made available by the oxidative steps of the cycle is transferred as energy-rich electrons to NAD^+, forming NADH. For each acetyl group that enters the citric acid cycle, three molecules of NADH are produced.

- Electrons are also transferred to the electron acceptor Q, forming QH_2.

- At the end of each cycle, the four-carbon oxaloacetate has been regenerated, and the cycle continues.

Steps

Two carbon atoms are oxidized to CO_2, the energy from these reactions being transferred to other metabolic processes by GTP (or ATP), and as electrons in NADH and QH_2. The NADH generated in the TCA cycle may later donate its electrons oxidative phosphorylation to drive ATP synthesis; $FADH_2$ is covalently attached to succinate dehydrogenase, an enzyme functioning both in the TCA cycle and the mitochondrial electron transport chain in oxidative phosphorylation. $FADH_2$, therefore, facilitates transfer of electrons to coenzyme Q, which is the final electron acceptor of the reaction catalyzed by the Succinate:ubiquinone oxidoreductase complex, also acting as an intermediate in the electron transport chain.

The citric acid cycle is continuously supplied with new carbon in the form of acetyl-CoA, entering at step 0 below.

	Substrates	Products	Enzyme	Reaction type	Comment
0 / 10	Oxaloacetate + Acetyl CoA + H_2O	Citrate + CoA-SH	Citrate synthase	Aldol condensation	irreversible, extends the 4C oxaloacetate to a 6C molecule
1	Citrate	cis-Aconitate + H_2O	Aconitase	Dehydration	reversible isomerisation
2	cis-Aconitate + H_2O	Isocitrate		Hydration	
3	Isocitrate + NAD^+	Oxalosuccinate + $NADH + H^+$	Isocitrate dehydrogenase	Oxidation	generates NADH (equivalent of 2.5 ATP)
4	Oxalosuccinate	α-Ketoglutarate + CO_2		Decarboxylation	rate-limiting, irreversible stage, generates a 5C molecule
5	α-Ketoglutarate + NAD^+ + CoA-SH	Succinyl-CoA + $NADH + H^+$ + CO_2	α-Ketoglutarate dehydrogenase	Oxidative decarboxylation	irreversible stage, generates NADH (equivalent of 2.5 ATP), regenerates the 4C chain (CoA excluded)
6	Succinyl-CoA + $GDP + P_i$	Succinate + CoA-SH + GTP	Succinyl-CoA synthetase	substrate-level phosphorylation	or ADP→ATP instead of GDP→GTP, generates 1 ATP or equivalent. Condensation reaction of GDP + P_i and hydrolysis of Succinyl-CoA involve the H_2O needed for balanced equation.
7	Succinate + ubiquinone (Q)	Fumarate + ubiquinol (QH_2)	Succinate dehydrogenase	Oxidation	uses FAD as a prosthetic group (FAD→$FADH_2$ in the first step of the reaction) in the enzyme, generates the equivalent of 1.5 ATP
8	Fumarate + H_2O	L-Malate	Fumarase	Hydration	Hydration of C-C double bond
9	L-Malate + NAD^+	Oxaloacetate + $NADH + H^+$	Malate dehydrogenase	Oxidation	reversible (in fact, equilibrium favors malate), generates NADH (equivalent of 2.5 ATP)

10 / 0	Oxaloacetate + Acetyl CoA + H_2O	Citrate + CoA-SH	Citrate synthase	Aldol condensation	This is the same as step 0 and restarts the cycle. The reaction is irreversible and extends the 4C oxaloacetate to a 6C molecule

Mitochondria in animals, including humans, possess two succinyl-CoA synthetases: one that produces GTP from GDP, and another that produces ATP from ADP. Plants have the type that produces ATP (ADP-forming succinyl-CoA synthetase). Several of the enzymes in the cycle may be loosely associated in a multienzyme protein complex within the mitochondrial matrix.

The GTP that is formed by GDP-forming succinyl-CoA synthetase may be utilized by nucleoside-diphosphate kinase to form ATP (the catalyzed reaction is GTP + ADP → GDP + ATP).

Products

Products of the first turn of the cycle are: *one GTP (or ATP), three NADH, one QH_2, two CO_2*.

Because two acetyl-CoA molecules are produced from each glucose molecule, two cycles are required per glucose molecule. Therefore, at the end of two cycles, the products are: two GTP, six NADH, two QH_2, and four CO_2

Description	Reactants	Products
The sum of all reactions in the citric acid cycle is:	Acetyl-CoA + 3 NAD^+ + Q + GDP + P_i + 2 H_2O	→ CoA-SH + 3 NADH + 3 H^+ + QH_2 + GTP + 2 CO_2
Combining the reactions occurring during the pyruvate oxidation with those occurring during the citric acid cycle, the following overall pyruvate oxidation reaction is obtained:	Pyruvate ion + 4 NAD^+ + Q + GDP + P_i + 2 H_2O	→ 4 NADH + 4 H^+ + QH_2 + GTP + 3 CO_2
Combining the above reaction with the ones occurring in the course of glycolysis, the following overall glucose oxidation reaction (excluding reactions in the respiratory chain) is obtained:	Glucose + 10 NAD^+ + 2 Q + 2 ADP + 2 GDP + 4 P_i + 2 H_2O	→ 10 NADH + 10 H^+ + 2 QH_2 + 2 ATP + 2 GTP + 6 CO_2

The above reactions are balanced if P_i represents the $H_2PO_4^-$ ion, ADP and GDP the ADP^{2-} and GDP^{2-} ions, respectively, and ATP and GTP the ATP^{3-} and GTP^{3-} ions, respectively.

The total number of ATP obtained after complete oxidation of one glucose in glycolysis, citric acid cycle, and oxidative phosphorylation is estimated to be between 30 and 38.

Efficiency

The theoretical maximum yield of ATP through oxidation of one molecule of glucose in glycolysis, citric acid cycle, and oxidative phosphorylation is 38 (assuming 3 molar equivalents of ATP per equivalent NADH and 2 ATP per $FADH_2$). In eukaryotes, two equivalents of NADH are generated in glycolysis, which takes place in the cytoplasm. Transport of these two equivalents into the mitochondria consumes two equivalents of ATP, thus reducing the net production of ATP to 36. Furthermore, inefficiencies in oxidative phosphorylation due to leakage of protons across the mitochondrial membrane and slippage of the ATP synthase/proton pump commonly reduces the ATP yield from NADH and $FADH_2$ to less than the theoretical maximum yield. The observed yields are, therefore, closer to ~2.5 ATP per NADH and ~1.5 ATP per $FADH_2$, further reducing the total net production of ATP to approximately 30. An assessment of the total ATP yield with newly revised proton-to-ATP ratios provides an estimate of 29.85 ATP per glucose molecule.

Variation

While the TCA cycle is in general highly conserved, there is significant variability in the enzymes found in different taxa (note that the diagrams on this page are specific to the mammalian pathway variant).

Some differences exist between eukaryotes and prokaryotes. The conversion of D-*threo*-isocitrate to 2-oxoglutarate is catalyzed in eukaryotes by the NAD^+-dependent EC 1.1.1.41, while prokaryotes employ the $NADP^+$-dependent EC 1.1.1.42. Similarly, the conversion of (*S*)-malate to oxaloacetate is catalyzed in eukaryotes by the NAD^+-dependent EC 1.1.1.37, while most prokaryotes utilize a quinone-dependent enzyme, EC 1.1.5.4.

A step with significant variability is the conversion of succinyl-CoA to succinate. Most organisms utilize EC 6.2.1.5, succinate–CoA ligase (ADP-forming) (despite its name, the enzyme operates in the pathway in the direction of ATP formation). In mammals a GTP-forming enzyme, succinate–CoA ligase (GDP-forming) (EC 6.2.1.4) also operates. The level of utilization of each isoform is tissue dependent. In some acetate-producing bacteria, such as *Acetobacter aceti*, an entirely different enzyme catalyzes this conversion – EC 2.8.3.18, succinyl-CoA:acetate CoA-transferase. This specialized enzyme links the TCA cycle with acetate metabolism in these organisms. Some bacteria, such as *Helicobacter pylori*, employ yet another enzyme for this conversion – succinyl-CoA:acetoacetate CoA-transferase (EC 2.8.3.5).

Some variability also exists at the previous step – the conversion of 2-oxoglutarate to succinyl-CoA. While most organisms utilize the ubiquitous NAD^+-dependent 2-oxoglutarate dehydrogenase, some bacteria utilize a ferredoxin-dependent 2-oxoglutarate synthase (EC 1.2.7.3). Other organisms, including obligately autotrophic and methanotrophic bacteria and archaea, bypass succinyl-CoA entirely, and convert 2-oxoglutarate to succinate via succinate semialdehyde, using EC 4.1.1.71, 2-oxoglutarate decarboxylase, and EC 1.2.1.79, succinate-semialdehyde dehydrogenase.

Regulation

The regulation of the TCA cycle is largely determined by product inhibition and substrate availability. If the cycle were permitted to run unchecked, large amounts of metabolic energy could be wasted in overproduction of reduced coenzyme such as NADH and ATP. The major eventual substrate of the cycle is ADP which gets converted to ATP. A reduced amount of ADP causes accumulation of precursor NADH which in turn can inhibit a number of enzymes. NADH, a product of all dehydrogenases in the TCA cycle with the exception of succinate dehydrogenase, inhibits pyruvate dehydrogenase, isocitrate dehydrogenase, α-ketoglutarate dehydrogenase, and also citrate synthase. Acetyl-coA inhibits pyruvate dehydrogenase, while succinyl-CoA inhibits alpha-ketoglutarate dehydrogenase and citrate synthase. When tested in vitro with TCA enzymes, ATP inhibits citrate synthase and α-ketoglutarate dehydrogenase; however, ATP levels do not change more than 10% in vivo between rest and vigorous exercise. There is no known allosteric mechanism that can account for large changes in reaction rate from an allosteric effector whose concentration changes less than 10%.

Calcium is used as a regulator. Mitochondrial matrix calcium levels can reach the tens of micromolar levels during cellular activation. It activates pyruvate dehydrogenase phosphatase which in turn activates the pyruvate dehydrogenase complex. Calcium also activates isocitrate dehydrogenase and α-ketoglutarate dehydrogenase. This increases the reaction rate of many of the steps in the cycle, and therefore increases flux throughout the pathway.

Citrate is used for feedback inhibition, as it inhibits phosphofructokinase, an enzyme involved in glycolysis that catalyses formation of fructose 1,6-bisphosphate,a precursor of pyruvate. This prevents a constant high rate of flux when there is an accumulation of citrate and a decrease in substrate for the enzyme.

Recent work has demonstrated an important link between intermediates of the citric acid cycle and the regulation of hypoxia-inducible factors (HIF). HIF plays a role in the regulation of oxygen homeostasis, and is a transcription factor that targets angiogenesis, vascular remodeling, glucose utilization, iron transport and apoptosis. HIF is synthesized consititutively, and hydroxylation of at least one of two critical proline residues mediates their interaction with the von Hippel Lindau E3 ubiquitin ligase complex, which targets them for rapid degradation. This reaction is catalysed by prolyl 4-hydroxylases. Fumarate and succinate have been identified as potent inhibitors of prolyl hydroxylases, thus leading to the stabilisation of HIF.

Major Metabolic Pathways Converging on The TCA Cycle

Several catabolic pathways converge on the TCA cycle. Most of these reactions add intermediates to the TCA cycle, and are therefore known as anaplerotic reactions, from the Greek meaning to "fill up". These increase the amount of acetyl CoA that the cycle is able to carry, increasing the mitochondrion's capability to carry out respiration if this is otherwise a limiting factor. Processes that remove intermediates from the cycle are termed "cataplerotic" reactions.

In this section and in the next, the citric acid cycle intermediates are indicated in *italics* to distinguish them from other substrates and end-products.

Pyruvate molecules produced by glycolysis are actively transported across the inner mitochondrial membrane, and into the matrix. Here they can be oxidized and combined with coenzyme A to form CO_2, *acetyl-CoA*, and NADH, as in the normal cycle.

However, it is also possible for pyruvate to be carboxylated by pyruvate carboxylase to form *oxaloacetate*. This latter reaction "fills up" the amount of *oxaloacetate* in the citric acid cycle, and is therefore an anaplerotic reaction, increasing the cycle's capacity to metabolize *acetyl-CoA* when the tissue's energy needs (e.g. in muscle) are suddenly increased by activity.

In the citric acid cycle all the intermediates (e.g. *citrate, iso-citrate, alpha-ketoglutarate, succinate, fumarate, malate* and *oxaloacetate*) are regenerated during each turn of the cycle. Adding more of any of these intermediates to the mitochondrion therefore means that that additional amount is retained within the cycle, increasing all the other intermediates as one is converted into the other. Hence the addition of any one of them to the cycle has an anaplerotic effect, and its removal has a cataplerotic effect. These anaplerotic and cataplerotic reactions will, during the course of the cycle, increase or decrease the amount of *oxaloacetate* available to combine with *acetyl-CoA* to form *citric acid*. This in turn increases or decreases the rate of ATP production by the mitochondrion, and thus the availability of ATP to the cell.

Acetyl-CoA, on the other hand, derived from pyruvate oxidation, or from the beta-oxidation of fatty acids, is the only fuel to enter the citric acid cycle. With each turn of the cycle one molecule of *acetyl-CoA* is consumed for every molecule of *oxaloacetate* present in the mitochondrial matrix, and is never regenerated. It is the oxidation of the acetate portion of *acetyl-CoA* that produces CO_2 and water, with the energy thus released captured in the form of ATP.

In the liver, the carboxylation of cytosolic pyruvate into intra-mitochondrial *oxaloacetate* is an early step in the gluconeogenic pathway which converts lactate and de-aminated alanine into glucose, under the influence of high levels of glucagon and/or epinephrine in the blood. Here the addition of *oxaloacetate* to the mitochondrion does not have a net anaplerotic effect, as another citric acid cycle intermediate (*malate*) is immediately removed from the mitochondrion to be converted into cytosolic oxaloace-tate, which is ultimately converted into glucose, in a process that is almost the reverse of glycolysis.

In protein catabolism, proteins are broken down by proteases into their constituent amino acids. Their carbon skeletons (i.e. the de-aminated amino acids) may either en-ter the citric acid cycle as intermediates (e.g. *alpha-ketoglutarate* derived from glu-tamate or glutamine), having an anaplerotic effect on the cycle, or, in the case of leu-cine, isoleucine, lysine, phenylalanine, tryptophan, and tyrosine, they are converted into *acetyl-CoA* which can be burned to CO_2 and water, or used to form ketone bodies, which too can only be burned in tissues other than the liver where they are formed, or excreted via the urine or breath. These latter amino acids are therefore termed "keto-genic" amino acids, whereas those that enter the citric acid cycle as intermediates can only be cataplerotically removed by entering the gluconeogenic pathway via *malate* which is transported out of the mitochondrion to be converted into cytosolic oxalo-acetate and ultimately into glucose. These are the so-called "glucogenic" amino acids. De-aminated alanine, cysteine, glycine, serine, and threonine are converted to pyruvate and can consequently either enter the citric acid cycle as *oxaloacetate* (an anaplerotic reaction) or as *acetyl-CoA* to be disposed of as CO_2 and water.

In fat catabolism, triglycerides are hydrolyzed to break them into fatty acids and glyc-erol. In the liver the glycerol can be converted into glucose via dihydroxyacetone phos-phate and glyceraldehyde-3-phosphate by way of gluconeogenesis. In many tissues, es-pecially heart and skeletal muscle tissue, fatty acids are broken down through a process known as beta oxidation, which results in the production of mitochondrial *acetyl-CoA*, which can be used in the citric acid cycle. Beta oxidation of fatty acids with an odd number of methylene bridges produces propionyl-CoA, which is then converted into *succinyl-CoA* and fed into the citric acid cycle as an anaplerotic intermediate.

The total energy gained from the complete breakdown of one (six-carbon) molecule of glucose by glycolysis, the formation of 2 *acetyl-CoA* molecules, their catabolism in the citric acid cycle, and oxidative phosphorylation equals about 30 ATP molecules, in eu-karyotes. The number of ATP molecules derived from the beta oxidation of a 6 carbon segment of a fatty acid chain, and the subsequent oxidation of the resulting 3 molecules of *acetyl-CoA* is 40.

Citric Acid Cycle Intermediates Serve as Substrates for Biosynthetic Processes

In this subheading, as in the previous one, the TCA intermediates are identified by *italics*.

Several of the citric acid cycle intermediates are used for the synthesis of important compounds, which will have significant cataplerotic effects on the cycle. *Acetyl-CoA* cannot be transported out of the mitochondrion. To obtain cytosolic acetyl-CoA, *citrate* is removed from the citric acid cycle and carried across the inner mitochondrial membrane into the cytosol. There it is cleaved by ATP citrate lyase into acetyl-CoA and oxaloacetate. The oxaloacetate is returned to mitochondrion as *malate* (and then converted back into *oxaloacetate* to transfer more *acetyl-CoA* out of the mitochondrion). The cytosolic acetyl-CoA is used for fatty acid synthesis and the production of cholesterol. Cholesterol can, in turn, be used to synthesize the steroid hormones, bile salts, and vitamin D.

The carbon skeletons of many non-essential amino acids are made from citric acid cycle intermediates. To turn them into amino acids the alpha keto-acids formed from the citric acid cycle intermediates have to acquire their amino groups from glutamate in a transamination reaction, in which pyridoxal phosphate is a cofactor. In this reaction the glutamate is converted into *alpha-ketoglutarate*, which is a citric acid cycle intermediate. The intermediates that can provide the carbon skeletons for amino acid synthesis are *oxaloacetate* which forms aspartate and asparagine; and *alpha-ketoglutarate* which forms glutamine, proline, and arginine.

Of these amino acids, aspartate and glutamine are used, together with carbon and nitrogen atoms from other sources, to form the purines that are used as the bases in DNA and RNA, as well as in ATP, AMP, GTP, NAD, FAD and CoA.

The pyrimidines are partly assembled from aspartate (derived from *oxaloacetate*). The pyrimidines, thymine, cytosine and uracil, form the complementary bases to the purine bases in DNA and RNA, and are also components of CTP, UMP, UDP and UTP.

The majority of the carbon atoms in the porphyrins come from the citric acid cycle intermediate, *succinyl-CoA*. These molecules are an important component of the hemoproteins, such as hemoglobin, myoglobin and various cytochromes.

During gluconeogenesis mitochondrial *oxaloacetate* is reduced to *malate* which is then transported out of the mitochondrion, to be oxidized back to oxaloacetate in the cytosol. Cytosolic oxaloacetate is then decarboxylated to phosphoenolpyruvate by phosphoenolpyruvate carboxykinase, which is the rate limiting step in the conversion of nearly all the gluconeogenic precursors (such as the glucogenic amino acids and lactate) into glucose by the liver and kidney.

Because the citric acid cycle is involved in both catabolic and anabolic processes, it is known as an amphibole pathway.

Citrate Synthase

The enzyme citrate synthase [E.C. 2.3.3.1 (previously 4.1.3.7)] exists in nearly all living cells and stands as a pace-making enzyme in the first step of the citric acid cycle (or Krebs cycle). Citrate synthase is localized within eukaryotic cells in the mitochondrial matrix, but is encoded by nuclear DNA rather than mitochondrial. It is synthesized using cytoplasmic ribosomes, then transported into the mitochondrial matrix. Citrate synthase is commonly used as a quantitative enzyme marker for the presence of intact mitochondria.

Citrate synthase catalyzes the condensation reaction of the two-carbon acetate residue from acetyl coenzyme A and a molecule of four-carbon oxaloacetate to form the six-carbon citrate:

- acetyl-CoA + oxaloacetate + H_2O → citrate + CoA-SH

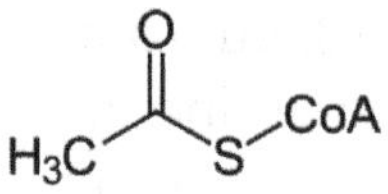

acetyl-CoA

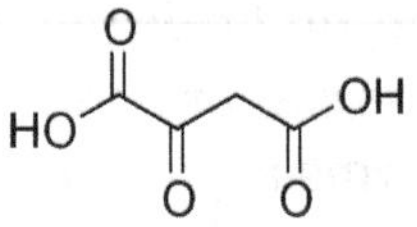

Oxaloacetic acid

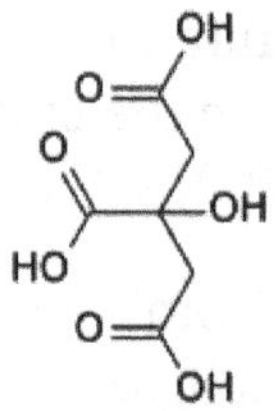

Citric acid

Oxaloacetate is regenerated after the completion of one round of the Krebs cycle.

Oxaloacetate is the first substrate to bind to the enzyme. This induces the enzyme to change its conformation, and creates a binding site for the acetyl-CoA. Only when this citroyl-CoA has formed will another conformational change cause thioester hydrolysis and release coenzyme A. This ensures that the energy released from the thioester bond cleavage will drive the condensation.

Structure

The Active Site of Citrate Synthase (open form)

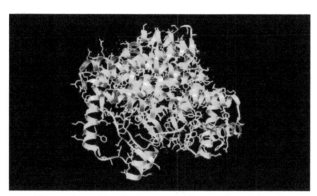

The Active Site of Citrate Synthase (closed form)

Citrate synthase's 437 amino acid residues are organized into two main subunits, each consisting of 20 alpha-helices. These alpha helices compose approximately 75% of citrate synthase's tertiary structure, while the remaining residues mainly compose irregular extensions of the structure, save a single beta-sheet of 13 residues. Between these two subunits, a single cleft exists containing the active site. Two binding sites can be found therein: one reserved for citrate or oxaloacetate and the other for Coenzyme A. The active site contains three key residues: His274, His320, and Asp375 that are highly selective in their interactions with substrates. The images to the left display the tertiary structure of citrate synthase in its opened and closed form. The enzyme changes from opened to closed with the addition of one of its substrates (such as oxaloacetate).

Function

Mechanism

Citrate synthase has three key amino acids in its active site (known as the catalytic triad) which catalyze the conversion of acetyl-CoA [$H_3CC(=O)-SCoA$] and oxaloacetate [$^-O_2CCH_2C(=O)CO_2^-$] into citrate [$^-O_2CCH_2C(OH)(CO_2^-)CH_2CO_2^-$] and H$-$SCoA in an aldol condensation reaction. This conversion begins with the negatively charged carboxylate side chain oxygen atom of Asp-375 deprotonating acetyl CoA's alpha carbon atom to form an enolate anion which in turn is neutralized by protonation by His-274 to form an enol intermediate [$H_2C=C(OH)-SCoA$]. At this point, the epsilon nitrogen lone pair of electrons on His-274 formed in the last step abstracts the hydroxyl enol proton to reform an enolate anion that initiates a nucleophilic attack on the oxaloacetate's carbonyl carbon [$^-O_2CCH_2C(=O)CO_2^-$] which in turn deprotonate the epsilon nitrogen atom of His-320. This nucleophilic addition results in the formation of citroyl$-$CoA [$^-O_2CCH_2CH(CO_2^-)CH_2C(=O)-SCoA$]. At this point, a water molecule is deprotonated by the epsilon nitrogen atom of His-320 and hydrolysis is initiated. One of the oxygen's lone pairs nucleophilically attacks the carbonyl carbon of citroyl$-$CoA. This forms a tetrahedral intermediate and results in the ejection of $-$SCoA as the carbonyl reforms. The $-$SCoA is protonated to form HSCoA. Finally, the hydroxyl added to the carbonyl in the previous step is deprotonated and citrate [$^-O_2CCH_2C(OH)(CO_2^-)CH_2CO_2^-$] is formed.

Mechanism for Citrate Synthase (including residues involved)

Inhibition

The enzyme is inhibited by high ratios of ATP:ADP, acetyl-CoA:CoA, and NADH:NAD, as high concentrations of ATP, acetyl-CoA, and NADH show that the energy supply is high for the cell. It is also inhibited by succinyl-CoA, which resembles Acetyl-coA and acts as a competitive inhibitor. Citrate inhibits the reaction and is an example of product inhibition. The inhibition of citrate synthase by acetyl-CoA analogues has also been well documented and has been used to prove the existence of a single active site. These experiments have revealed that this single site alternates between two forms, which participate in ligase and hydrolase activity respectively. This protein may use the morpheein model of allosteric regulation.

Oxidative Phosphorylation

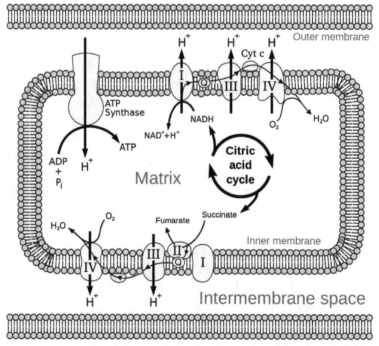

The electron transport chain in the cell is the site of oxidative phosphorylation in prokaryotes. The NADH and succinate generated in the citric acid cycle are oxidized, releasing energy to power the ATP synthase.

Oxidative phosphorylation (or OXPHOS in short) is the metabolic pathway in which cells use enzymes to oxidize nutrients, thereby releasing energy which is used to re-form ATP. In most eukaryotes, this takes place inside mitochondria. Almost all aerobic organisms carry out oxidative phosphorylation. This pathway is probably so pervasive because it is a highly efficient way of releasing energy, compared to alternative fermentation processes such as anaerobic glycolysis.

During oxidative phosphorylation, electrons are transferred from electron donors to electron acceptors such as oxygen, in redox reactions. These redox reactions release energy, which is used to form ATP. In eukaryotes, these redox reactions are carried out by a series of protein complexes within the inner membrane of the cell's mitochondria, whereas, in prokaryotes, these proteins are located in the cells' intermembrane space. These linked sets of proteins are called electron transport chains. In eukaryotes, five main protein complexes are involved, whereas in prokaryotes many different enzymes are present, using a variety of electron donors and acceptors.

The energy released by electrons flowing through this electron transport chain is used to transport protons across the inner mitochondrial membrane, in a process called *electron transport*. This generates potential energy in the form of a pH gradient and an electrical potential across this membrane. This store of energy is tapped by allowing protons to flow back across the membrane and down this gradient, through a large enzyme called ATP synthase; this process is known as chemiosmosis. This enzyme uses this energy to generate ATP from adenosine diphosphate (ADP), in a phosphorylation reaction. This reaction is driven by the proton flow, which forces the rotation of a part of the enzyme; the ATP synthase is a rotary mechanical motor.

Although oxidative phosphorylation is a vital part of metabolism, it produces reactive oxygen species such as superoxide and hydrogen peroxide, which lead to propagation of free radicals, damaging cells and contributing to disease and, possibly, aging (senescence). The enzymes carrying out this metabolic pathway are also the target of many drugs and poisons that inhibit their activities.

Overview of Energy Transfer by Chemiosmosis

Oxidative phosphorylation works by using energy-releasing chemical reactions to drive energy-requiring reactions: The two sets of reactions are said to be *coupled*. This means one cannot occur without the other. The flow of electrons through the electron transport chain, from electron donors such as NADH to electron acceptors such as oxygen, is an exergonic process – it releases energy, whereas the synthesis of ATP is an endergonic process, which requires an input of energy. Both the electron transport chain and the ATP synthase are embedded in a membrane, and energy is transferred from electron transport chain to the ATP synthase by movements of protons across this membrane, in a process called *chemiosmosis*. In practice, this is like a simple electric circuit, with a current of protons being driven from the negative N-side of the membrane to the positive P-side by the proton-pumping enzymes of the electron transport chain. These enzymes are like a battery, as they perform work to drive current through the circuit. The movement of protons creates an electrochemical gradient across the membrane, which is often called the *proton-motive force*. It has two components: a difference in proton concentration (a H^+ gradient, ΔpH) and a difference in electric potential, with the N-side having a negative charge.

ATP synthase releases this stored energy by completing the circuit and allowing protons to flow down the electrochemical gradient, back to the N-side of the membrane. This kinetic energy drives the rotation of part of the enzymes structure and couples this motion to the synthesis of ATP.

The two components of the proton-motive force are thermodynamically equivalent: In mitochondria, the largest part of energy is provided by the potential; in alkaliphile bacteria the electrical energy even has to compensate for a counteracting inverse pH difference. Inversely, chloroplasts operate mainly on ΔpH. However, they also require a small membrane potential for the kinetics of ATP synthesis. At least in the case of the fusobacterium *P. modestum* it drives the counter-rotation of subunits a and c of the F_0 motor of ATP synthase.

The amount of energy released by oxidative phosphorylation is high, compared with the amount produced by anaerobic fermentation. Glycolysis produces only 2 ATP molecules, but somewhere between 30 and 36 ATPs are produced by the oxidative phosphorylation of the 10 NADH and 2 succinate molecules made by converting one molecule of glucose to carbon dioxide and water, while each cycle of beta oxidation of a fatty acid yields about 14 ATPs. These ATP yields are theoretical maximum values; in practice, some protons leak across the membrane, lowering the yield of ATP.

Electron and Proton Transfer Molecules

$$2\,e^- + 2\,H^+$$

Reduction of coenzyme Q from its ubiquinone form (Q) to the reduced ubiquinol form (QH_2).

The electron transport chain carries both protons and electrons, passing electrons from donors to acceptors, and transporting protons across a membrane. These processes use both soluble and protein-bound transfer molecules. In mitochondria, electrons are transferred within the intermembrane space by the water-soluble electron transfer protein cytochrome c. This carries only electrons, and these are transferred by the reduction and oxidation of an iron atom that the protein holds within a heme group in its structure. Cytochrome c is also found in some bacteria, where it is located within the periplasmic space.

Within the inner mitochondrial membrane, the lipid-soluble electron carrier coenzyme Q10 (Q) carries both electrons and protons by a redox cycle. This small benzoquinone molecule is very hydrophobic, so it diffuses freely within the membrane. When Q accepts two electrons and two protons, it becomes reduced to the *ubiquinol* form (QH_2); when QH_2 releases two electrons and two protons, it becomes oxidized back to the *ubiquinone* (Q) form. As a result, if two enzymes are arranged so that Q is reduced on one side of the membrane and QH_2 oxidized on the other, ubiquinone will couple these reactions and shuttle protons across the membrane. Some bacterial electron transport chains use different quinones, such as menaquinone, in addition to ubiquinone.

Within proteins, electrons are transferred between flavin cofactors, iron–sulfur clusters, and cytochromes. There are several types of iron–sulfur cluster. The simplest kind found in the electron transfer chain consists of two iron atoms joined by two atoms of inorganic sulfur; these are called [2Fe–2S] clusters. The second kind, called [4Fe–4S], contains a cube of four iron atoms and four sulfur atoms. Each iron atom in these clusters is coordinated by an additional amino acid, usually by the sulfur atom of cysteine. Metal ion cofactors undergo redox reactions without binding or releasing protons, so in the electron transport chain they serve solely to transport electrons through proteins. Electrons move quite long distances through proteins by hopping along chains of these cofactors. This occurs by quantum tunnelling, which is rapid over distances of less than 1.4×10^{-9} m.

Eukaryotic Electron Transport Chains

Many catabolic biochemical processes, such as glycolysis, the citric acid cycle, and beta oxidation, produce the reduced coenzyme NADH. This coenzyme contains electrons that have a high transfer potential; in other words, they will release a large amount of energy upon oxidation. However, the cell does not release this energy all at once, as this would be an uncontrollable reaction. Instead, the electrons are removed from NADH and passed to oxygen through a series of enzymes that each release a small amount of the energy. This set of enzymes, consisting of complexes I through IV, is called the electron transport chain and is found in the inner membrane of the mitochondrion. Succinate is also oxidized by the electron transport chain, but feeds into the pathway at a different point.

In eukaryotes, the enzymes in this electron transport system use the energy released from the oxidation of NADH to pump protons across the inner membrane of the mitochondrion.

This causes protons to build up in the intermembrane space, and generates an electro-chemical gradient across the membrane. The energy stored in this potential is then used by ATP synthase to produce ATP. Oxidative phosphorylation in the eukaryotic mitochondrion is the best-understood example of this process. The mitochondrion is present in almost all eukaryotes, with the exception of anaerobic protozoa such as *Trichomonas vaginalis* that instead reduce protons to hydrogen in a remnant mitochondrion called a hydrogenosome.

Typical respiratory enzymes and substrates in eukaryotes.		
Respiratory enzyme	Redox pair	Midpoint potential (Volts)
NADH dehydrogenase	NAD^+ / NADH	−0.32
Succinate dehydrogenase	FMN or FAD / $FMNH_2$ or $FADH_2$	−0.20
Cytochrome bc_1 complex	Coenzyme Q10$_{ox}$ / Coenzyme Q10$_{red}$	+0.06
Cytochrome bc_1 complex	Cytochrome b_{ox} / Cytochrome b_{red}	+0.12
Complex IV	Cytochrome c_{ox} / Cytochrome c_{red}	+0.22
Complex IV	Cytochrome a_{ox} / Cytochrome a_{red}	+0.29
Complex IV	O_2 / HO^-	+0.82
Conditions: pH = 7		

Nadh-Coenzyme Q Oxidoreductase (Complex I)

NADH-coenzyme Q oxidoreductase, also known as *NADH dehydrogenase* or *complex I*, is the first protein in the electron transport chain. Complex I is a giant enzyme with the mammalian complex I having 46 subunits and a molecular mass of about 1,000 kilodaltons (kDa). The structure is known in detail only from a bacterium; in most organisms the complex resembles a boot with a large "ball" poking out from the membrane into the mitochondrion. The genes that encode the individual proteins are contained in both the cell nucleus and the mitochondrial genome, as is the case for many enzymes present in the mitochondrion.

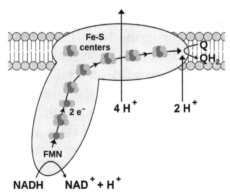

Complex I or NADH-Q oxidoreductase. The abbreviations are discussed in the text. In all diagrams of respiratory complexes in this article, the matrix is at the bottom, with the intermembrane space above.

The reaction that is catalyzed by this enzyme is the two electron oxidation of NADH by coenzyme Q10 or *ubiquinone* (represented as Q in the equation below), a lipid-soluble quinone that is found in the mitochondrion membrane:

$$NADH + Q + 5H^+_{matrix} \rightarrow NAD^+ + QH_2 + 4H^+_{intermembrane} \tag{1}$$

The start of the reaction, and indeed of the entire electron chain, is the binding of a NADH molecule to complex I and the donation of two electrons. The electrons enter complex I via a prosthetic group attached to the complex, flavin mononucleotide (FMN). The addition of electrons to FMN converts it to its reduced form, $FMNH_2$. The electrons are then transferred through a series of iron–sulfur clusters: the second kind of prosthetic group present in the complex. There are both [2Fe–2S] and [4Fe–4S] iron–sulfur clusters in complex I.

As the electrons pass through this complex, four protons are pumped from the matrix into the intermembrane space. Exactly how this occurs is unclear, but it seems to involve conformational changes in complex I that cause the protein to bind protons on the N-side of the membrane and release them on the P-side of the membrane. Finally, the electrons are transferred from the chain of iron–sulfur clusters to a ubiquinone molecule in the membrane. Reduction of ubiquinone also contributes to the generation of a proton gradient, as two protons are taken up from the matrix as it is reduced to ubiquinol (QH_2).

Succinate-Q Oxidoreductase (Complex Ii)

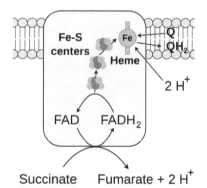

Complex II: Succinate-Q oxidoreductase.

Succinate-Q oxidoreductase, also known as *complex II* or *succinate dehydrogenase*, is a second entry point to the electron transport chain. It is unusual because it is the only enzyme that is part of both the citric acid cycle and the electron transport chain. Complex II consists of four protein subunits and contains a bound flavin adenine dinucleotide (FAD) cofactor, iron–sulfur clusters, and a heme group that does not participate in electron transfer to coenzyme Q, but is believed to be important in decreasing production of reactive oxygen species. It oxidizes succinate to fumarate and reduces ubiquinone. As this reaction releases less energy than the oxidation of NADH, complex

II does not transport protons across the membrane and does not contribute to the proton gradient.

$$Succinate + Q \rightarrow Fumarate + QH_2 \tag{2}$$

In some eukaryotes, such as the parasitic worm *Ascaris suum*, an enzyme similar to complex II, fumarate reductase (menaquinol:fumarate oxidoreductase, or QFR), operates in reverse to oxidize ubiquinol and reduce fumarate. This allows the worm to survive in the anaerobic environment of the large intestine, carrying out anaerobic oxidative phosphorylation with fumarate as the electron acceptor. Another unconventional function of complex II is seen in the malaria parasite *Plasmodium falciparum*. Here, the reversed action of complex II as an oxidase is important in regenerating ubiquinol, which the parasite uses in an unusual form of pyrimidine biosynthesis.

Electron Transfer Flavoprotein-Q Oxidoreductase

Electron transfer flavoprotein-ubiquinone oxidoreductase (ETF-Q oxidoreductase), also known as *electron transferring-flavoprotein dehydrogenase*, is a third entry point to the electron transport chain. It is an enzyme that accepts electrons from electron-transferring flavoprotein in the mitochondrial matrix, and uses these electrons to reduce ubiquinone. This enzyme contains a flavin and a [4Fe–4S] cluster, but, unlike the other respiratory complexes, it attaches to the surface of the membrane and does not cross the lipid bilayer.

$$ETF_{red} + Q \rightarrow ETF_{ox} + QH_2 \tag{3}$$

In mammals, this metabolic pathway is important in beta oxidation of fatty acids and catabolism of amino acids and choline, as it accepts electrons from multiple acetyl-CoA dehydrogenases. In plants, ETF-Q oxidoreductase is also important in the metabolic responses that allow survival in extended periods of darkness.

Q-Cytochrome C Oxidoreductase (Complex III)

Q-cytochrome c oxidoreductase is also known as *cytochrome c reductase, cytochrome bc_1 complex*, or simply *complex III*. In mammals, this enzyme is a dimer, with each subunit complex containing 11 protein subunits, an [2Fe-2S] iron–sulfur cluster and three cytochromes: one cytochrome c_1 and two b cytochromes. A cytochrome is a kind of electron-transferring protein that contains at least one heme group. The iron atoms inside complex III's heme groups alternate between a reduced ferrous (+2) and oxidized ferric (+3) state as the electrons are transferred through the protein.

The reaction catalyzed by complex III is the oxidation of one molecule of ubiquinol and the reduction of two molecules of cytochrome c, a heme protein loosely associated with the mitochondrion. Unlike coenzyme Q, which carries two electrons, cytochrome c carries only one electron.

As only one of the electrons can be transferred from the QH_2 donor to a cytochrome c acceptor at a time, the reaction mechanism of complex III is more elaborate than those of the other respiratory complexes, and occurs in two steps called the Q cycle. In the first step, the enzyme binds three substrates, first, QH_2, which is then oxidized, with one electron being passed to the second substrate, cytochrome c. The two protons released from QH_2 pass into the intermembrane space. The third substrate is Q, which accepts the second electron from the QH_2 and is reduced to $Q^{\cdot-}$, which is the ubisemiquinone free radical. The first two substrates are released, but this ubisemiquinone intermediate remains bound. In the second step, a second molecule of QH_2 is bound and again passes its first electron to a cytochrome c acceptor. The second electron is passed to the bound ubisemiquinone, reducing it to QH_2 as it gains two protons from the mitochondrial matrix. This QH_2 is then released from the enzyme.

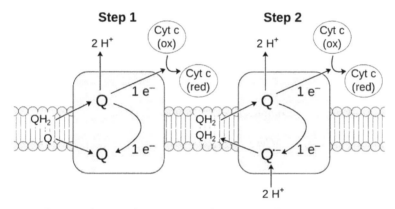

The two electron transfer steps in complex III: Q-cytochrome c oxidoreductase. After each step, Q (in the upper part of the figure) leaves the enzyme.

$$QH_2 + 2Cytc_{ox} + 2H^+{}_{matrix} \rightarrow Q + 2Cytc_{red} + 4H^+{}_{intermembrane} \qquad (4)$$

As coenzyme Q is reduced to ubiquinol on the inner side of the membrane and oxidized to ubiquinone on the other, a net transfer of protons across the membrane occurs, adding to the proton gradient. The rather complex two-step mechanism by which this occurs is important, as it increases the efficiency of proton transfer. If, instead of the Q cycle, one molecule of QH_2 were used to directly reduce two molecules of cytochrome c, the efficiency would be halved, with only one proton transferred per cytochrome c reduced.

Cytochrome C Oxidase (Complex IV)

Cytochrome c oxidase, also known as *complex IV*, is the final protein complex in the electron transport chain. The mammalian enzyme has an extremely complicated structure and contains 13 subunits, two heme groups, as well as multiple metal ion cofactors – in all, three atoms of copper, one of magnesium and one of zinc.

This enzyme mediates the final reaction in the electron transport chain and transfers electrons to oxygen, while pumping protons across the membrane. The final electron acceptor oxygen, which is also called the *terminal electron acceptor*, is reduced to water in this step. Both the direct pumping of protons and the consumption of matrix protons in the reduction of oxygen contribute to the proton gradient. The reaction catalyzed is the oxidation of cytochrome c and the reduction of oxygen:

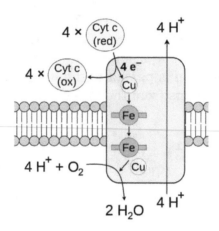

Complex IV: cytochrome c oxidase.

$$4Cytc_{red} + O_2 + 8H^+_{matrix} \rightarrow 4Cytc_{ox} + 2H_2O + 4H^+_{intermembrane} \tag{5}$$

Alternative Reductases and Oxidases

Many eukaryotic organisms have electron transport chains that differ from the much-studied mammalian enzymes described above. For example, plants have alternative NADH oxidases, which oxidize NADH in the cytosol rather than in the mitochondrial matrix, and pass these electrons to the ubiquinone pool. These enzymes do not transport protons, and, therefore, reduce ubiquinone without altering the electrochemical gradient across the inner membrane.

Another example of a divergent electron transport chain is the *alternative oxidase*, which is found in plants, as well as some fungi, protists, and possibly some animals. This enzyme transfers electrons directly from ubiquinol to oxygen.

The electron transport pathways produced by these alternative NADH and ubiquinone oxidases have lower ATP yields than the full pathway. The advantages produced by a shortened pathway are not entirely clear. However, the alternative oxidase is produced in response to stresses such as cold, reactive oxygen species, and infection by pathogens, as well as other factors that inhibit the full electron transport chain. Alternative pathways might, therefore, enhance an organisms' resistance to injury, by reducing oxidative stress.

Organization of Complexes

The original model for how the respiratory chain complexes are organized was that they diffuse freely and independently in the mitochondrial membrane. However, recent data suggest that the complexes might form higher-order structures called supercomplexes or "respirasomes." In this model, the various complexes exist as organized sets of interacting enzymes. These associations might allow channeling of substrates between the various enzyme complexes, increasing the rate and efficiency of electron transfer. Within such mammalian supercomplexes, some components would be present in higher amounts than others, with some data suggesting a ratio between complexes I/II/III/IV and the ATP synthase of approximately 1:1:3:7:4. However, the debate over this supercomplex hypothesis is not completely resolved, as some data do not appear to fit with this model.

Prokaryotic Electron Transport Chains

In contrast to the general similarity in structure and function of the electron transport chains in eukaryotes, bacteria and archaea possess a large variety of electron-transfer enzymes. These use an equally wide set of chemicals as substrates. In common with eukaryotes, prokaryotic electron transport uses the energy released from the oxidation of a substrate to pump ions across a membrane and generate an electrochemical gradient. In the bacteria, oxidative phosphorylation in *Escherichia coli* is understood in most detail, while archaeal systems are at present poorly understood.

The main difference between eukaryotic and prokaryotic oxidative phosphorylation is that bacteria and archaea use many different substances to donate or accept electrons. This allows prokaryotes to grow under a wide variety of environmental conditions. In *E. coli*, for example, oxidative phosphorylation can be driven by a large number of pairs of reducing agents and oxidizing agents, which are listed below. The midpoint potential of a chemical measures how much energy is released when it is oxidized or reduced, with reducing agents having negative potentials and oxidizing agents positive potentials.

Respiratory enzymes and substrates in *E. coli*.		
Respiratory enzyme	Redox pair	Midpoint potential (Volts)
Formate dehydrogenase	Bicarbonate / Formate	−0.43
Hydrogenase	Proton / Hydrogen	−0.42
NADH dehydrogenase	NAD+ / NADH	−0.32
Glycerol-3-phosphate dehydrogenase	DHAP / Gly-3-P	−0.19
Pyruvate oxidase	Acetate + Carbon dioxide / Pyruvate	?
Lactate dehydrogenase	Pyruvate / Lactate	−0.19
D-amino acid dehydrogenase	2-oxoacid + ammonia / D-amino acid	?
Glucose dehydrogenase	Gluconate / Glucose	−0.14

Succinate dehydrogenase	Fumarate / Succinate	+0.03
Ubiquinol oxidase	Oxygen / Water	+0.82
Nitrate reductase	Nitrate / Nitrite	+0.42
Nitrite reductase	Nitrite / Ammonia	+0.36
Dimethyl sulfoxide reductase	DMSO / DMS	+0.16
Trimethylamine N-oxide reductase	TMAO / TMA	+0.13
Fumarate reductase	Fumarate / Succinate	+0.03

As shown above, *E. coli* can grow with reducing agents such as formate, hydrogen, or lactate as electron donors, and nitrate, DMSO, or oxygen as acceptors. The larger the difference in midpoint potential between an oxidizing and reducing agent, the more energy is released when they react. Out of these compounds, the succinate/fumarate pair is unusual, as its midpoint potential is close to zero. Succinate can therefore be oxidized to fumarate if a strong oxidizing agent such as oxygen is available, or fumarate can be reduced to succinate using a strong reducing agent such as formate. These alternative reactions are catalyzed by succinate dehydrogenase and fumarate reductase, respectively.

Some prokaryotes use redox pairs that have only a small difference in midpoint potential. For example, nitrifying bacteria such as *Nitrobacter* oxidize nitrite to nitrate, donating the electrons to oxygen. The small amount of energy released in this reaction is enough to pump protons and generate ATP, but not enough to produce NADH or NADPH directly for use in anabolism. This problem is solved by using a nitrite oxidoreductase to produce enough proton-motive force to run part of the electron transport chain in reverse, causing complex I to generate NADH.

Prokaryotes control their use of these electron donors and acceptors by varying which enzymes are produced, in response to environmental conditions. This flexibility is possible because different oxidases and reductases use the same ubiquinone pool. This allows many combinations of enzymes to function together, linked by the common ubiquinol intermediate. These respiratory chains therefore have a modular design, with easily interchangeable sets of enzyme systems.

In addition to this metabolic diversity, prokaryotes also possess a range of isozymes – different enzymes that catalyze the same reaction. For example, in *E. coli*, there are two different types of ubiquinol oxidase using oxygen as an electron acceptor. Under highly aerobic conditions, the cell uses an oxidase with a low affinity for oxygen that can transport two protons per electron. However, if levels of oxygen fall, they switch to an oxidase that transfers only one proton per electron, but has a high affinity for oxygen.

ATP Synthase (Complex V)

ATP synthase, also called *complex V*, is the final enzyme in the oxidative phosphorylation pathway. This enzyme is found in all forms of life and functions in the same way in both prokaryotes and eukaryotes. The enzyme uses the energy stored in a proton gradient across a membrane to drive the synthesis of ATP from ADP and phosphate (P_i). Estimates of the number of protons required to synthesize one ATP have ranged from three to four, with some suggesting cells can vary this ratio, to suit different conditions.

$$ADP + P_i + 4H^+{}_{intermembrane} \rightleftharpoons ATP + H_2O + 4H^+{}_{matrix} \qquad (6)$$

This phosphorylation reaction is an equilibrium, which can be shifted by altering the proton-motive force. In the absence of a proton-motive force, the ATP synthase reaction will run from right to left, hydrolyzing ATP and pumping protons out of the matrix across the membrane. However, when the proton-motive force is high, the reaction is forced to run in the opposite direction; it proceeds from left to right, allowing protons to flow down their concentration gradient and turning ADP into ATP. Indeed, in the closely related vacuolar type H+-ATPases, the hydrolysis reaction is used to acidify cellular compartments, by pumping protons and hydrolysing ATP.

ATP synthase is a massive protein complex with a mushroom-like shape. The mammalian enzyme complex contains 16 subunits and has a mass of approximately 600 kilodaltons. The portion embedded within the membrane is called F_0 and contains a ring of c subunits and the proton channel. The stalk and the ball-shaped headpiece is called F_1 and is the site of ATP synthesis. The ball-shaped complex at the end of the F_1 portion contains six proteins of two different kinds (three α subunits and three β subunits), whereas the "stalk" consists of one protein: the γ subunit, with the tip of the stalk extending into the ball of α and β subunits. Both the α and β subunits bind nucleotides, but only the β subunits catalyze the ATP synthesis reaction. Reaching along the side of the F_1 portion and back into the membrane is a long rod-like subunit that anchors the α and β subunits into the base of the enzyme.

As protons cross the membrane through the channel in the base of ATP synthase, the F_0 proton-driven motor rotates. Rotation might be caused by changes in the ionization of amino acids in the ring of c subunits causing electrostatic interactions that propel the ring of c subunits past the proton channel. This rotating ring in turn drives the rotation of the central axle (the γ subunit stalk) within the α and β subunits. The α and β subunits are prevented from rotating themselves by the side-arm, which acts as a stator. This movement of the tip of the γ subunit within the ball of α and β subunits provides the energy for the active sites in the β subunits to undergo a cycle of movements that produces and then releases ATP.

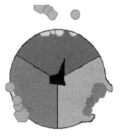

Mechanism of ATP synthase. ATP is shown in red, ADP and phosphate in pink and the rotating γ subunit in black.

This ATP synthesis reaction is called the *binding change mechanism* and involves the active site of a β subunit cycling between three states. In the "open" state, ADP and phosphate enter the active site (shown in brown in the diagram). The protein then closes up around the molecules and binds them loosely – the "loose" state (shown in red). The enzyme then changes shape again and forces these molecules together, with the active site in the resulting "tight" state (shown in pink) binding the newly produced ATP molecule with very high affinity. Finally, the active site cycles back to the open state, releasing ATP and binding more ADP and phosphate, ready for the next cycle.

In some bacteria and archaea, ATP synthesis is driven by the movement of sodium ions through the cell membrane, rather than the movement of protons. Archaea such as *Methanococcus* also contain the A_1A_0 synthase, a form of the enzyme that contains additional proteins with little similarity in sequence to other bacterial and eukaryotic ATP synthase subunits. It is possible that, in some species, the A_1A_0 form of the enzyme is a specialized sodium-driven ATP synthase, but this might not be true in all cases.

Reactive Oxygen Species

Molecular oxygen is an ideal terminal electron acceptor because it is a strong oxidizing agent. The reduction of oxygen does involve potentially harmful intermediates. Although the transfer of four electrons and four protons reduces oxygen to water, which is harmless, transfer of one or two electrons produces superoxide or peroxide anions, which are dangerously reactive.

$$O_2 \xrightarrow{e^-} \underset{\text{Superoxide}}{O_2^{\bullet-}} \xrightarrow{e^-} \underset{\text{Peroxide}}{O_2^{2-}} \tag{7}$$

These reactive oxygen species and their reaction products, such as the hydroxyl radical, are very harmful to cells, as they oxidize proteins and cause mutations in DNA. This cellular damage might contribute to disease and is proposed as one cause of aging.

The cytochrome c oxidase complex is highly efficient at reducing oxygen to water, and it releases very few partly reduced intermediates; however small amounts of superoxide anion and peroxide are produced by the electron transport chain. Particularly important is the reduction of coenzyme Q in complex III, as a highly reactive ubisemiquinone free radical is formed as an intermediate in the Q cycle. This unstable species can lead to electron "leakage" when electrons transfer directly to oxygen, forming superoxide. As the production of reactive oxygen species by these proton-pumping complexes is greatest at high membrane potentials, it has been proposed that mitochondria regulate their activity to maintain the membrane potential within a narrow range that balances ATP production against oxidant generation. For instance, oxidants can activate uncoupling proteins that reduce membrane potential.

To counteract these reactive oxygen species, cells contain numerous antioxidant systems, including antioxidant vitamins such as vitamin C and vitamin E, and antioxidant enzymes such as superoxide dismutase, catalase, and peroxidases, which detoxify the reactive species, limiting damage to the cell.

Inhibitors

There are several well-known drugs and toxins that inhibit oxidative phosphorylation. Although any one of these toxins inhibits only one enzyme in the electron transport chain, inhibition of any step in this process will halt the rest of the process. For example, if oligomycin inhibits ATP synthase, protons cannot pass back into the mitochondrion. As a result, the proton pumps are unable to operate, as the gradient becomes too strong for them to overcome. NADH is then no longer oxidized and the citric acid cycle ceases to operate because the concentration of NAD^+ falls below the concentration that these enzymes can use.

Compounds	Use	Site of action	Effect on oxidative phosphorylation
Cyanide Carbon monoxide Azide Hydrogen sulfide	Poisons	Complex IV	Inhibit the electron transport chain by binding more strongly than oxygen to the Fe–Cu center in cytochrome c oxidase, preventing the reduction of oxygen.
Oligomycin	Antibiotic	Complex V	Inhibits ATP synthase by blocking the flow of protons through the F_o subunit.
CCCP 2,4-Dinitrophenol	Poisons, weight-loss[N 1]	Inner membrane	Ionophores that disrupt the proton gradient by carrying protons across a membrane. This ionophore uncouples proton pumping from ATP synthesis because it carries protons across the inner mitochondrial membrane.
Rotenone	Pesticide	Complex I	Prevents the transfer of electrons from complex I to ubiquinone by blocking the ubiquinone-binding site.
Malonate and oxaloacetate	Poisons	Complex II	Competitive inhibitors of succinate dehydrogenase (complex II).
Antimycin A	Piscicide	Complex III	Binds to the Qi site of cytochrome c reductase, thereby inhibiting the oxidation of ubiquinol.

Not all inhibitors of oxidative phosphorylation are toxins. In brown adipose tissue, regulated proton channels called uncoupling proteins can uncouple respiration from ATP synthesis. This rapid respiration produces heat, and is particularly important as a way of maintaining body temperature for hibernating animals, although these proteins may also have a more general function in cells' responses to stress.

History

The field of oxidative phosphorylation began with the report in 1906 by Arthur Harden of a vital role for phosphate in cellular fermentation, but initially only sugar phosphates were known to be involved. However, in the early 1940s, the link between the oxidation of sugars and the generation of ATP was firmly established by Herman Kalckar, confirming the central role of ATP in energy transfer that had been proposed by Fritz Albert Lipmann in 1941. Later, in 1949, Morris Friedkin and Albert L. Lehninger proved that the coenzyme NADH linked metabolic pathways such as the citric acid cycle and the synthesis of ATP. The term *oxidative phosphorylation* was coined by Volodymyr Belitser (uk) in 1939.

For another twenty years, the mechanism by which ATP is generated remained mysterious, with scientists searching for an elusive "high-energy intermediate" that would link oxidation and phosphorylation reactions. This puzzle was solved by Peter D. Mitchell with the publication of the chemiosmotic theory in 1961. At first, this proposal was highly controversial, but it was slowly accepted and Mitchell was awarded a Nobel prize in 1978. Subsequent research concentrated on purifying and characterizing the enzymes involved, with major contributions being made by David E. Green on the complexes of the electron-transport chain, as well as Efraim Racker on the ATP synthase. A critical step towards solving the mechanism of the ATP synthase was provided by Paul D. Boyer, by his development in 1973 of the "binding change" mechanism, followed by his radical proposal of rotational catalysis in 1982. More recent work has included structural studies on the enzymes involved in oxidative phosphorylation by John E. Walker, with Walker and Boyer being awarded a Nobel Prize in 1997.

Glycolysis

Glycolysis (from *glycose*, an older term for glucose + -*lysis* degradation) is the metabolic pathway that converts glucose $C_6H_{12}O_6$, into pyruvate, $CH_3COCOO^- + H^+$. The free energy released in this process is used to form the high-energy compounds ATP (adenosine triphosphate) and NADH (reduced nicotinamide adenine dinucleotide).

Glycolysis is a determined sequence of ten enzyme-catalyzed reactions. The intermediates provide entry points to glycolysis. For example, most monosaccharides, such as fructose and galactose, can be converted to one of these intermediates. The intermediates may also be directly useful. For example, the intermediate dihydroxyacetone phosphate (DHAP) is a source of the glycerol that combines with fatty acids to form fat.

Glycolysis is an oxygen independent metabolic pathway, meaning that it does not use molecular oxygen (i.e. atmospheric oxygen) for any of its reactions. However the products of glycolysis (pyruvate and NADH + H^+) are sometimes disposed of using atmospheric oxygen. When molecular oxygen is used in the disposal of the products of glycolysis the process is usually referred to as aerobic, whereas if the disposal uses no oxygen the process is said to be anaerobic. Thus, glycolysis occurs, with variations, in nearly all organisms, both aerobic and anaerobic. The wide occurrence of glycolysis indicates that it is one of the most ancient metabolic pathways. Indeed, the reactions that constitute glycolysis and its parallel pathway, the pentose phosphate pathway, occur metal-catalyzed under the oxygen-free conditions of the Archean oceans, also in the absence of enzymes. Glycolysis could thus have originated from chemical constraints of the prebiotic world.

Glycolysis occurs in most organisms in the cytosol of the cell. The most common type of glycolysis is the *Embden–Meyerhof–Parnas (EMP pathway)*, which was discovered by Gustav Embden, Otto Meyerhof, and Jakub Karol Parnas. Glycolysis also refers to other pathways, such as the *Entner–Doudoroff pathway* and various heterofermentative and homofermentative pathways. However, the discussion here will be limited to the Embden–Meyerhof–Parnas pathway.

Overview

The use of symbols in this equation makes it appear not balanced with respect to oxygen atoms, hydrogen atoms, and charges. Atom balance is maintained by the two phosphate (P_i) groups:

- Each exists in the form of a hydrogen phosphate anion (HPO_4^{2-}), dissociating to contribute 2 H^+ overall
- Each liberates an oxygen atom when it binds to an ADP (adenosine diphosphate) molecule, contributing 2 O overall

Charges are balanced by the difference between ADP and ATP. In the cellular environment, all three hydroxyl groups of ADP dissociate into $-O^-$ and H^+, giving ADP^{3-}, and this ion tends to exist in an ionic bond with Mg^{2+}, giving $ADPMg^-$. ATP behaves identically except that it has four hydroxyl groups, giving $ATPMg^{2-}$. When these differences along with the true charges on the two phosphate groups are considered together, the net charges of -4 on each side are balanced.

For simple fermentations, the metabolism of one molecule of glucose to two molecules of pyruvate has a net yield of two molecules of ATP. Most cells will then carry out further reactions to 'repay' the used NAD^+ and produce a final product of ethanol or lactic acid. Many bacteria use inorganic compounds as hydrogen acceptors to regenerate the NAD^+.

Cells performing aerobic respiration synthesize much more ATP, but not as part of glycolysis. These further aerobic reactions use pyruvate and NADH + H^+ from glycolysis. Eukaryotic aerobic respiration produces approximately 34 additional molecules of ATP for each glucose molecule, however most of these are produced by a vastly different mechanism to the substrate-level phosphorylation in glycolysis.

The lower-energy production, per glucose, of anaerobic respiration relative to aerobic respiration, results in greater flux through the pathway under hypoxic (low-oxygen) conditions, unless alternative sources of anaerobically oxidizable substrates, such as fatty acids, are found.

History

The pathway of glycolysis as it is known today took almost 100 years to fully discover. The combined results of many smaller experiments were required in order to understand the pathway as a whole.

The first steps in understanding glycolysis began in the nineteenth century with the wine industry. For economic reasons, the French wine industry sought to investigate why wine sometime turned distasteful, instead of fermenting into alcohol. French scientist Louis Pasteur researched this issue during the 1850s, and the results of his experiments began the long road to elucidating the pathway of glycolysis. His experiments showed that fermentation occurs by the action of living microorganisms; and that yeast's glucose consumption decreased under aerobic conditions of fermentation, in comparison to anaerobic conditions (the Pasteur Effect).

Eduard Buchner. Discovered cell-free fermentation.

While Pasteur's experiments were groundbreaking, insight into the component steps of glycolysis were provided by the non-cellular fermentation experiments of Eduard Buchner during the 1890s. Buchner demonstrated that the conversion of glucose to ethanol was possible using a non-living extract of yeast (due to the action of enzymes in the extract). This experiment not only revolutionized biochemistry, but also allowed later scientists to analyze this pathway in a more controlled lab setting. In a series of experiments (1905-1911), scientists Arthur Harden and William Young discovered more pieces of glycolysis. . They discovered the regulatory effects of ATP on glucose consumption during alcohol fermentation. They also shed light on the role of one compound as a glycolysis intermediate: fructose 1,6-bisphosphate.

The elucidation of Fructose 1,6-diphosphate was accomplished by measuring CO_2 levels when yeast juice was incubated with glucose. CO_2 production increased rapidly then slowed down. Harden and Young noted that this process would restart if an inorganic phosphate (Pi) was added to the mixture. Harden and Young deduced that this process produced organic phosphate esters, and further experiments allowed them to extract fructose diphosphate (F-1,6-DP).

Arthur Harden and William Young along with Nick Sheppard determined, in a second experiment, that a heat-sensitive high-molecular-weight subcellular fraction (the enzymes) and a heat-insensitive low-molecular-weight cytoplasm fraction (ADP, ATP and NAD$^+$ and other cofactors) are required together for fermentation to proceed. This experiment begun by observing that dialyzed (purified) yeast juice could not ferment or even create a sugar phosphate. This mixture was rescued with the addition of undialyzed yeast extract that had been boiled. Boiling the yeast extract renders all proteins inactive (as it denatures them). The ability of boiled extract plus dialyzed juice to complete fermentation suggests that the cofactors were non-protein in character.

Otto Meyerhof. One of the main scientists involved in completing the puzzle of glycolysis

In the 1920s Otto Meyerhof was able to link together some of the many individual pieces of glycolysis discovered by Buchner, Harden, and Young. Meyerhof and his team was able to extract different glycolytic enzymes from muscle tissue, and combine them to artificially create the pathway from glycogen to lactic acid.

In one paper, Meyerhof and scientist Renate Junowicz-Kockolaty investigated the reaction that splits fructose 1,6-diphosohate into the two triose phosphates. Previous work proposed that the split occurred via 1,3-diphosphoglyceraldehye plus an oxidizing enzyme and cozymase. Meyerhoff and Junowicz found that the equilibrium constant for the isomerase and aldoses reaction were not affected by inorganic phosphates or any other cozymase or oxidizing enzymes. They further removed diphosphoglyceraldehyde as a possible intermediate in glycolysis.

With all of these pieces available by the 1930s, Gustav Embden proposed a detailed, step-by-step outline of that pathway we now know as glycolysis. The biggest difficulties in determining the intricacies of the pathway were due to the very short lifetime and low steady-state concentrations of the intermediates of the fast glycolytic reactions. By the 1940s, Meyerhof, Embden and many other biochemists had finally completed the puzzle of glycolysis. The understanding of the isolated pathway has been expanded in the subsequent decades, to include further details of its regulation and integration with other metabolic pathways.

Regulation

Glycolysis is regulated by slowing down or speeding up certain steps in the pathway by inhibiting or activating the enzymes that are involved. The steps that are regulated may be determined by calculating the change in free energy, ΔG, for each step.

When ΔG is negative, a reaction proceeds spontaneously in the forward direction only and is considered irreversible. When ΔG is positive, the reaction is non-spontaneous and will not proceed in the forward direction unless coupled with an energetically favorable reaction. When ΔG is zero, the reaction is at equilibrium, can proceed in either directions and is considered reversible.

If a step is at equilibrium (ΔG is zero), the enzyme catalyzing the reaction will balance the products and reactants and cannot confer directionality to the pathway. These steps (and associated enzymes) are considered unregulated. If a step is not at equilibrium, but spontaneous (ΔG is negative), the enzyme catalyzing the reaction is not balancing the products and reactants and is considered to be regulated. A common mechanism of regulating enzymes is allosteric control.

Free Energy Changes

Concentrations of metabolites in erythrocytes	
Compound	Concentration / mM
Glucose	5.0
Glucose-6-phosphate	0.083
Fructose-6-phosphate	0.014
Fructose-1,6-bisphosphate	0.031
Dihydroxyacetone phosphate	0.14
Glyceraldehyde-3-phosphate	0.019
1,3-Bisphosphoglycerate	0.001
2,3-Bisphosphoglycerate	4.0
3-Phosphoglycerate	0.12
2-Phosphoglycerate	0.03
Phosphoenolpyruvate	0.023
Pyruvate	0.051
ATP	1.85

ADP	0.14
P_i	1.0

The change in free energy, ΔG, for each step in the glycolysis pathway can be calculated using $\Delta G = \Delta G^{o'} + RT\ln Q$, where Q is the reaction quotient. This requires knowing the concentrations of the metabolites. All of these values are available for erythrocytes, with the exception of the concentrations of NAD^+ and NADH. The ratio of NAD^+ to NADH in the cytoplasm is approximately 1000, which makes the oxidation of glyceraldehyde-3-phosphate (step 6) more favourable.

Using the measured concentrations of each step, and the standard free energy changes, the actual free energy change can be calculated. (Neglecting this is very common - the delta G of ATP hydrolysis in cells is not the standard free energy change of ATP hydrolysis quoted in textbooks).

Change in free energy for each step of glycolysis			
Step	Reaction	$\Delta G^{o'}$ / (kJ/ mol)	ΔG / (kJ/ mol)
1	Glucose + $ATP^{4-} \rightarrow$ Glucose-6-phosphate^{2-} + ADP^{3-} + H^+	-16.7	-34
2	Glucose-6-phosphate$^{2-} \rightarrow$ Fructose-6-phosphate^{2-}	1.67	-2.9
3	Fructose-6-phosphate^{2-} + $ATP^{4-} \rightarrow$ Fructose-1,6-bisphosphate^{4-} + ADP^{3-} + H^+	-14.2	-19
4	Fructose-1,6-bisphosphate$^{4-} \rightarrow$ Dihydroxyacetone phosphate^{2-} + Glyceraldehyde-3-phosphate^{2-}	23.9	-0.23
5	Dihydroxyacetone phosphate$^{2-} \rightarrow$ Glyceraldehyde-3-phosphate^{2-}	7.56	2.4
6	Glyceraldehyde-3-phosphate^{2-} + P_i^{2-} + $NAD^+ \rightarrow$ 1,3-Bisphosphoglycerate^{4-} + NADH + H^+	6.30	-1.29
7	1,3-Bisphosphoglycerate^{4-} + $ADP^{3-} \rightarrow$ 3-Phosphoglycerate^{3-} + ATP^{4-}	-18.9	0.09
8	3-Phosphoglycerate$^{3-} \rightarrow$ 2-Phosphoglycerate^{3-}	4.4	0.83
9	2-Phosphoglycerate$^{3-} \rightarrow$ Phosphoenolpyruvate^{3-} + H_2O	1.8	1.1
10	Phosphoenolpyruvate^{3-} + ADP^{3-} + $H^+ \rightarrow$ Pyruvate$^-$ + ATP^{4-}	-31.7	-23.0

From measuring the physiological concentrations of metabolites in an erythrocyte it seems that about seven of the steps in glycolysis are in equilibrium for that cell type. Three of the steps — the ones with large negative free energy changes — are not in equilibrium and are referred to as *irreversible*; such steps are often subject to regulation.

Step 5 in the figure is shown behind the other steps, because that step is a side-reaction that can decrease or increase the concentration of the intermediate glyceraldehyde-3-phosphate. That compound is converted to dihydroxyacetone phosphate by the enzyme triose phosphate isomerase, which is a catalytically perfect enzyme; its rate is so fast that the reaction can be assumed to be in equilibrium. The fact that ΔG is not zero indicates that the actual concentrations in the erythrocyte are not accurately known.

Biochemical Logic

The existence of more than one point of regulation indicates that intermediates between those points enter and leave the glycolysis pathway by other processes. For example, in the first regulated step, hexokinase converts glucose into glucose-6-phosphate. Instead of continuing through the glycolysis pathway, this intermediate can be converted into glucose storage molecules, such as glycogen or starch. The reverse reaction, breaking down, e.g., glycogen, produces mainly glucose-6-phosphate; very little free glucose is formed in the reaction. The glucose-6-phosphate so produced can enter glycolysis *after* the first control point.

In the second regulated step (the third step of glycolysis), phosphofructokinase converts fructose-6-phosphate into fructose-1,6-bisphosphate, which then is converted into glyceraldehyde-3-phosphate and dihydroxyacetone phosphate. The dihydroxyacetone phosphate can be removed from glycolysis by conversion into glycerol-3-phosphate, which can be used to form triglycerides. On the converse, triglycerides can be broken down into fatty acids and glycerol; the latter, in turn, can be converted into dihydroxyacetone phosphate, which can enter glycolysis *after* the second control point.

Regulation of The Rate Limiting Enzymes

The four regulatory enzymes are hexokinase, glucokinase, phosphofructokinase, and pyruvate kinase. The flux through the glycolytic pathway is adjusted in response to conditions both inside and outside the cell. The internal factors that regulate glycolysis do so primarily to provide ATP in adequate quantities for the cell's needs. The external factors act primarily on the liver, fat tissue, and muscles, which can remove large quantities of glucose from the blood after meals (thus preventing hyperglycemia by storing the excess glucose as fat or glycogen, depending on the tissue type). The liver is also capable of releasing glucose into the blood between meals, during fasting, and exercise thus preventing hypoglycemia by means of glycogenolysis and gluconeogenesis. These latter reactions coincide with the halting of glycolysis in the liver.

In animals, regulation of blood glucose levels by the pancreas in conjunction with the liver is a vital part of homeostasis. The beta cells in the pancreatic islets are sensitive to the blood glucose concentration. A rise in the blood glucose concentration causes them to release insulin into the blood, which has an effect particularly on the liver, but also on fat and muscle cells, causing these tissues to remove glucose from the blood. When the blood sugar falls the pancreatic beta cells cease insulin production, but, instead, stimulate the neighboring pancreatic alpha cells to release glucagon into the blood. This, in turn, causes the liver to release glucose into the blood by breaking down stored glycogen, and by means of gluconeogenesis. If the fall in the blood glucose level is particularly rapid or severe, other glucose sensors cause the release of epinephrine from the adrenal glands into the blood. This has the same action as glucagon on glucose metabolism, but its effect is more pronounced. In the liver glucagon and epinephrine cause the phosphorylation of the key, rate limiting enzymes of glycolysis, fatty acid synthesis, cholesterol synthesis, gluconeogenesis, and glycogenolysis. Insulin has the opposite effect on these enzymes. The phosphorylation and dephosphorylation of these enzymes (ultimately in response to the glucose level in the blood) is the dominant manner by which these pathways are controlled in the liver, fat, and muscle cells. Thus the phosphorylation of phosphofructokinase inhibits glycolysis, whereas its dephosphorylation through the action of insulin stimulates glycolysis.

In addition hexokinase and glucokinase act independently of the hormonal effects as controls at the entry points of glucose into the cells of different tissues. Hexokinase responds to the glucose-6-phosphate (G6P) level in the cell, or, in the case of glucokinase, to the blood sugar level in the blood to impart entirely intracellular controls of the glycolytic pathway in different tissues.

When glucose has been converted into G6P by hexokinase or glucokinase, it can either be converted to glucose-1-phosphate (G1P) for conversion to glycogen, or it is alternatively converted by glycolysis to pyruvate, which enters the mitochondrion where it is converted into acetyl-CoA and then into citrate. Excess citrate is exported from the mitochondrion back into to the cytosol, where ATP citrate lyase regenerates acetyl-CoA and oxaloacetate (OAA). The acetyl-CoA is then used for fatty acid synthesis and cholesterol synthesis, two important ways of utilizing excess glucose when its concentration is high in blood. The rate limiting enzymes catalyzing these reactions perform these functions when they have been dephosphorylated through the action of insulin on the liver cells. Between meals, during fasting, exercise or hypoglycemia, glucagon and epinephrine are released into the blood. This causes liver glycogen to be converted back to G6P, and then converted to glucose by the liver-specific enzyme glucose 6-phosphatase and released into the blood. Glucagon and epinephrine also stimulate gluconeogenesis, which coverts non-carbohydrate substrates into G6P, which joins the G6P derived from glycogen, or substitutes for it when the liver glycogen store have been depleted. This is critical for brain function, since the brain utilizes glucose as an energy source under most conditions. The simultaneously phosphorylation of, particularly, phosphofructokinase, but also, to a certain extent pyruvate kinase, prevents glycolysis occurring at the same time as gluconeogenesis and glycogenolysis.

Hexokinase and Glucokinase

Cytochrome c oxidase, also known as *complex IV*, is the final protein complex in the electron transport chain. The mammalian enzyme has an extremely complicated structure and contains 13 subunits, two heme groups, as well as multiple metal ion cofactors – in all, three atoms of copper, one of magnesium and one of zinc.

All cells contain the enzyme hexokinase, which catalyzes the conversion of glucose that has entered the cell into glucose-6-phosphate (G6P). Since the cell wall is impervious to G6P, hexokinase essentially acts to transport glucose into the cells from which it can then no longer escape. Hexokinase is inhibited by high levels of G6P in the cell. Thus the rate of entry of glucose into cells partially depends on how fast G6P can be disposed of by glycolysis, and by glycogen synthesis (in the cells which store glycogen, namely liver and muscles).

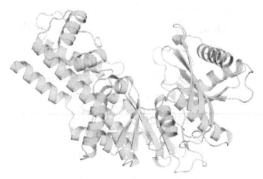

Yeast hexokinase B (PDB: 1IG8)

Glucokinase, unlike hexokinase, is not inhibited by G6P. It occurs in liver cells, and will only phosphorylate the glucose entering the cell to form glucose-6-phosphate (G6P), when the sugar in the blood is abundant. This being the first step in the glycolytic pathway in the liver, it therefore imparts an additional layer of control of the glycolytic pathway in this organ.

Phosphofructokinase

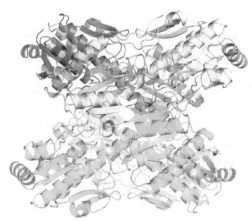

Bacillus stearothermophilus phosphofructokinase (PDB: 6PFK)

Phosphofructokinase is an important control point in the glycolytic pathway, since it is one of the irreversible steps and has key allosteric effectors, AMP and fructose 2,6-bisphosphate (F2,6BP).

Fructose 2,6-bisphosphate (F2,6BP) is a very potent activator of phosphofructokinase (PFK-1) that is synthesized when F6P is phosphorylated by a second phosphofructokinase (PFK2). In liver, when blood sugar is low and glucagon elevates cAMP, PFK2 is phosphorylated by protein kinase A. The phosphorylation inactivates PFK2, and another domain on this protein becomes active as fructose bisphosphatase-2, which converts F2,6BP back to F6P. Both glucagon and epinephrine cause high levels of cAMP in the liver. The result of lower levels of liver fructose-2,6-bisphosphate is a decrease in activity of phosphofructokinase and an increase in activity of fructose 1,6-bisphosphatase, so that gluconeogenesis (in essence, "glycolysis in reverse") is favored. This is consistent with the role of the liver in such situations, since the response of the liver to these hormones is to release glucose to the blood.

ATP competes with AMP for the allosteric effector site on the PFK enzyme. ATP concentrations in cells are much higher than those of AMP, typically 100-fold higher, but the concentration of ATP does not change more than about 10% under physiological conditions, whereas a 10% drop in ATP results in a 6-fold increase in AMP. Thus, the relevance of ATP as an allosteric effector is questionable. An increase in AMP is a consequence of a decrease in energy charge in the cell.

Citrate inhibits phosphofructokinase when tested *in vitro* by enhancing the inhibitory effect of ATP. However, it is doubtful that this is a meaningful effect *in vivo*, because citrate in the cytosol is utilized mainly for conversion to acetyl-CoA for fatty acid and cholesterol synthesis.

Pyruvate Kinase

Pyruvate kinase enzyme catalyzes the last step of glycolysis, in which pyruvate and ATP are formed. Pyruvate kinase catalyzes the transfer of a phosphate group from phosphoenolpyruvate (PEP) to ADP, yielding one molecule of pyruvate and one molecule of ATP.

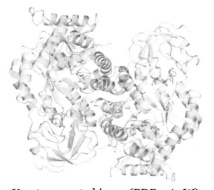

Yeast pyruvate kinase (PDB: 1A3W)

Liver pyruvate kinase is indirectly regulated by epinephrine and glucagon, through protein kinase A. This protein kinase phosphorylates liver pyruvate kinase to deactivate it. Muscle pyruvate kinase is not inhibited by epinephrine activation of protein kinase A.

Glucagon signals fasting (no glucose available). Thus, glycolysis is inhibited in the liver but unaffected in muscle when fasting. An increase in blood sugar leads to secretion of insulin, which activates phosphoprotein phosphatase I, leading to dephosphorylation and activation of pyruvate kinase. These controls prevent pyruvate kinase from being active at the same time as the enzymes that catalyze the reverse reaction (pyruvate carboxylase and phosphoenolpyruvate carboxykinase), preventing a futile cycle.

Post-Glycolysis Processes

The overall process of glycolysis is:

$$\text{Glucose} + 2\ NAD^+ + 2\ ADP + 2\ P_i \rightarrow 2\ \text{Pyruvate} + 2\ NADH + 2\ H^+ + 2\ ATP + 2\ H_2O$$

If glycolysis were to continue indefinitely, all of the NAD^+ would be used up, and glycolysis would stop. To allow glycolysis to continue, organisms must be able to oxidize NADH back to NAD^+. How this is performed depends on which external electron acceptor is available.

Anoxic Regeneration of NAD+

One method of doing this is to simply have the pyruvate do the oxidation; in this process, pyruvate is converted to lactate (the conjugate base of lactic acid) in a process called lactic acid fermentation:

$$\text{Pyruvate} + NADH + H^+ \rightarrow \text{Lactate} + NAD^+$$

This process occurs in the bacteria involved in making yogurt (the lactic acid causes the milk to curdle). This process also occurs in animals under hypoxic (or partially anaerobic) conditions, found, for example, in overworked muscles that are starved of oxygen. In many tissues, this is a cellular last resort for energy; most animal tissue cannot tolerate anaerobic conditions for an extended period of time.

Some organisms, such as yeast, convert NADH back to NAD^+ in a process called ethanol fermentation. In this process, the pyruvate is converted first to acetaldehyde and carbon dioxide, and then to ethanol.

Lactic acid fermentation and ethanol fermentation can occur in the absence of oxygen. This anaerobic fermentation allows many single-cell organisms to use glycolysis as their only energy source.

Anoxic regeneration of NAD^+ is only an effective means of energy production during short, intense exercise in vertebrates, for a period ranging from 10 seconds to 2 minutes during a maximal effort in humans. (At lower exercise intensities it can sustain muscle activity in diving animals, such as seals, whales and other aquatic vertebrates, for very much longer periods of time.) Under these conditions NAD^+ is replenished by NADH donating its electrons to pyruvate to form lactate. This produces 2 ATP molecules per glucose molecule, or about 5% of glucose's energy potential (38 ATP molecules in bacteria). But the speed at which ATP is produced in this manner is about 100 times that of oxidative phosphorylation. The pH in the cytoplasm quickly drops when hydrogen ions accumulate in the muscle, eventually inhibiting the enzymes involved in glycolysis.

The burning sensation in muscles during hard exercise can be attributed to the production of hydrogen ions during the shift to glucose fermentation from glucose oxidation to carbon dioxide and water, when aerobic metabolism can no longer keep pace with the energy demands of the muscles. These hydrogen ions form a part of lactic acid. The body falls back on this less efficient but faster method of producing ATP under low oxygen conditions. This is thought to have been the primary means of energy production in earlier organisms before oxygen reached high concentrations in the atmosphere between 2000 and 2500 million years ago, and thus would represent a more ancient form of energy production than the aerobic replenishment of NAD^+ in cells.

The liver in mammals gets rid of this excess lactate by transforming it back into pyruvate under aerobic conditions.

Fermenation of pyruvate to lactate is sometimes also called "anaerobic glycolysis", however, glycolysis ends with the production of pyruvate regardless of the presence or absence of oxygen.

In the above two examples of fermentation, NADH is oxidized by transferring two electrons to pyruvate. However, anaerobic bacteria use a wide variety of compounds as the terminal electron acceptors in cellular respiration: nitrogenous compounds, such as nitrates and nitrites; sulfur compounds, such as sulfates, sulfites, sulfur dioxide, and elemental sulfur; carbon dioxide; iron compounds; manganese compounds; cobalt compounds; and uranium compounds.

Aerobic Regeneration of NAD^+, and Disposal of Pyruvate

In aerobic organisms, a complex mechanism has been developed to use the oxygen in air as the final electron acceptor.

- Firstly, the NADH + H^+ generated by glycolysis has to be transferred to the mitochondrion to be oxidized, and thus to regenerate the NAD^+ necessary for glycolysis to continue. However the inner mitochondrial membrane is impermeable to NADH and NAD^+. Use is therefore made of two "shuttles" to transport the electrons from NADH across the mitochondrial membrane. They are the malate-aspartate shuttle and the glycerol phosphate shuttle. In the former the electrons from NADH are transferred to cytosolic oxaloacetate to form malate. The malate then traverses the inner mitochondrial membrane into the mitochondrial matrix, where it is reoxidized by NAD^+ forming intra-mitochondrial oxaloacetate and NADH. The oxaloacetate is then re-cycled to the cytosol via its conversion to aspartate which is readily transported out of the mitochondrion. In the glycerol phosphate shuttle electrons from cytosolic NADH are transferred to dihydroxyacetone to form glycerol-3-phosphate which readily traverses the outer mitochondrial membrane. Glycerol-3-phosphate is then reoxidized to di-hydroxyacetone, donating its electrons to FAD instead of NAD^+. This reaction takes place on the inner mitochondrial membrane, allowing $FADH_2$ to donate its electrons directly to coenzyme Q (ubiquinone) which is part of the electron transport chain which ultimately transfers electrons to molecular oxygen (O_2), with the formation of water, and the release of energy eventually captured in the form of ATP.

- The glycolytic end-product, pyruvate (plus NAD^+) is converted to acetyl-CoA, CO_2 and $NADH + H^+$ within the mitochondria in a process called pyruvate de-carboxylation.

- The resulting acetyl-CoA enters the citric acid cycle (or Krebs Cycle), where the acetyl group of the acetyl-CoA is converted into carbon dioxide by two decar-boxylation reactions with the formation of yet more intra-mitochondrial $NADH + H^+$.

- The intra-mitochondrial $NADH + H^+$ is oxidized to NAD^+ by the electron trans-port chain, using oxygen as the final electron acceptor to form water. The energy released during this process is used to create a hydrogen ion (or proton) gradi-ent across the inner membrane of the mitochondrion.

- Finally, the proton gradient is used to produce about 2.5 ATP for every $NADH + H^+$ oxidized in a process called oxidative phosphorylation.

Conversion of Carbohydrates into Fatty Acids and Cholesterol

The pyruvate produced by glycolysis is an important intermediary in the conversion of carbohydrates into fatty acids and cholesterol. This occurs via the conversion of py-ruvate into acetyl-CoA in the mitochondrion. However, this acetyl CoA needs to be transported into cytosol where the synthesis of fatty acids and cholesterol occurs. This cannot occur directly. To obtain cytosolic acetyl-CoA, citrate (produced by the con-densation of acetyl CoA with oxaloacetate) is removed from the citric acid cycle and carried across the inner mitochondrial membrane into the cytosol. There it is cleaved by ATP citrate lyase into acetyl-CoA and oxaloacetate. The oxaloacetate is returned to mitochondrion as malate (and then back into oxaloacetate to transfer more acetyl-CoA out of the mitochondrion). The cytosolic acetyl-CoA can be carboxylated by acetyl-CoA carboxylase into malonyl CoA, the first committed step in the synthesis of fatty acids, or it can be combined with acetoacetyl-CoA to form 3-hydroxy-3-methylglutaryl-CoA (HMG-CoA) which is the rate limiting step controlling the synthesis of cholesterol. Cholesterol can be used as is, as a structural component of cellular membranes, or it can be used to synthesize the steroid hormones, bile salts, and vitamin D.

Conversion of Pyruvate into Oxaloacetate for The Citric Acid Cycle

Pyruvate molecules produced by glycolysis are actively transported across the inner mitochondrial membrane, and into the matrix where they can either be oxidized and combined with coenzyme A to form CO_2, acetyl-CoA, and NADH, or they can be carboxylated (by pyruvate carboxylase) to form oxaloacetate. This latter reaction "fills up" the amount of oxaloacetate in the citric acid cycle, and is therefore an anaplerotic reaction (from the Greek meaning to "fill up"), increasing the cycle's capacity to metabolize acetyl-CoA when the tissue's energy needs (e.g. in heart and skeletal muscle) are suddenly increased by activity. In the citric acid cycle all the intermediates (e.g. citrate, iso-citrate, alpha-ketoglutarate, succinate, fumarate, malate and oxaloacetate) are regenerated during each turn of the cycle. Adding more of any of these intermediates to the mitochondrion therefore means that that additional amount is retained within the cycle, increasing all the other intermediates as one is converted into the other. Hence the addition of oxaloacetate greatly increases the amounts of all the citric acid intermediates, thereby increasing the cycle's capacity to metabolize acetyl CoA, converting its acetate component into CO_2 and water, with the release of enough energy to form 11 ATP and 1 GTP molecule for each additional molecule of acetyl CoA that combines with oxaloacetate in the cycle.

To cataplerotically remove oxaloacetate from the citric cycle, malate can be transported from the mitochondrion into the cytoplasm, decreasing the amount of oxaloacetate that can be regenerated. Furthermore, citric acid intermediates are constantly used to form a variety of substances such as the purines, pyrimidines and porphyrins.

Intermediates for Other Pathways

This article concentrates on the catabolic role of glycolysis with regard to converting potential chemical energy to usable chemical energy during the oxidation of glucose to pyruvate. Many of the metabolites in the glycolytic pathway are also used by anabolic pathways, and, as a consequence, flux through the pathway is critical to maintain a supply of carbon skeletons for biosynthesis.

The following metabolic pathways are all strongly reliant on glycolysis as a source of metabolites: and many more.

- Pentose phosphate pathway, which begins with the dehydrogenation of glucose-6-phosphate, the first intermediate to be produced by glycolysis, produces various pentose sugars, and NADPH for the synthesis of fatty acids and cholesterol.

- Glycogen synthesis also starts with glucose-6-phosphate at the beginning of the glycolytic pathway.

- Glycerol, for the formation of triglycerides and phospholipids, is produced from the glycolytic intermediate glyceraldehyde-3-phosphate.

- Various post-glycolytic pathways:

  - Fatty acid synthesis

  - Cholesterol synthesis

  - The citric acid cycle which in turn leads to:

  - Amino acid synthesis

  - Nucleotide synthesis

  - Tetrapyrrole synthesis

Although gluconeogenesis and glycolysis share many intermediates the one is not functionally a branch or tributary of the other. There are two regulatory steps in both pathways which, when active in the one pathway, are automatically inactive in the other. The two processes can therefore not be simultaneously active. Indeed, if both sets of reactions were highly active at the same time the net result would be the hydrolysis of four high energy phosphate bonds (two ATP and two GTP) per reaction cycle.

NAD^+ is the oxidizing agent in glycolysis, as it is in most other energy yielding metabolic reactions (e.g. beta-oxidation of fatty acids, and during the citric acid cycle). The NADH thus produced is primarily used to ultimately transfer electrons to O_2 to produce water, or, when O_2 is not available, to produced compounds such as lactate or ethanol. NADH is rarely used for synthetic processes, the notable exception being gluconeogenesis. During fatty acid and cholesterol synthesis the reducing agent is NADPH. This difference exemplifies a general principle that NADPH is consumed during biosynthetic reactions, whereas NADH is generated in energy-yielding reactions. The source of the NADPH is two-fold. When malate is oxidatively decarboxylated by "$NADP^+$-linked malic enzyme" pyruvate, CO_2 and NADPH are formed. NADPH is also formed by the pentose phosphate pathway which converts glucose into ribose, which can be used in synthesis of nucleotides and nucleic acids, or it can be catabolized to pyruvate.

Glycolysis in Disease

Genetic diseases

Glycolytic mutations are generally rare due to importance of the metabolic pathway, this means that the majority of occurring mutations result in an inability for the cell to respire, and therefore cause the death of the cell at an early stage. However, some mutations are seen with one notable example being Pyruvate kinase deficiency, leading to chronic hemolytic anemia.

Cancer

Malignant rapidly growing tumor cells typically have glycolytic rates that are up to 200 times higher than those of their normal tissues of origin. This phenomenon was first described in 1930 by Otto Warburg and is referred to as the Warburg effect. The Warburg hypothesis claims that cancer is primarily caused by dysfunctionality in mitochondrial metabolism, rather than because of uncontrolled growth of cells. A number of theories have been advanced to explain the Warburg effect. One such theory suggests that the increased glycolysis is a normal protective process of the body and that malignant change could be primarily caused by energy metabolism.

This high glycolysis rate has important medical applications, as high aerobic glycolysis by malignant tumors is utilized clinically to diagnose and monitor treatment responses of cancers by imaging uptake of $2\text{-}^{18}F\text{-}2$-deoxyglucose (FDG) (a radioactive modified hexokinase substrate) with positron emission tomography (PET).

There is ongoing research to affect mitochondrial metabolism and treat cancer by reducing glycolysis and thus starving cancerous cells in various new ways, including a ketogenic diet.

Alternative Nomenclature

Some of the metabolites in glycolysis have alternative names and nomenclature. In part, this is because some of them are common to other pathways, such as the Calvin cycle.

	This article		Alternative names	Alternative nomenclature
1	Glucose	Glc	Dextrose	
3	Fructose-6-phosphate	F6P		
4	Fructose-1,6-bisphosphate	F1,6BP	Fructose 1,6-diphosphate	FBP, FDP, F1,6DP
5	Dihydroxyacetone phosphate	DHAP	Glycerone phosphate	
6	Glyceraldehyde-3-phosphate	GADP	3-Phosphoglyceraldehyde	PGAL, G3P, GALP,GAP,TP
7	1,3-Bisphosphoglycerate	1,3BPG	Glycerate-1,3-bisphosphate, glycerate-1,3-diphosphate, 1,3-diphosphoglycerate	PGAP, BPG, DPG
8	3-Phosphoglycerate	3PG	Glycerate-3-phosphate	PGA, GP
9	2-Phosphoglycerate	2PG	Glycerate-2-phosphate	
10	Phosphoenolpyruvate	PEP		
11	Pyruvate	Pyr	Pyruvic acid	

Phosphofructokinase 1

Phosphofructokinase-1 (PFK-1) is one of the most important regulatory enzymes (EC 2.7.1.11) of glycolysis. It is an allosteric enzyme made of 4 subunits and controlled by many activators and inhibitors. PFK-1 catalyzes the important "committed" step of glycolysis, the conversion of fructose 6-phosphate and ATP to fructose 1,6-bisphosphate and ADP. Glycolysis is the foundation for respiration, both anaerobic and aerobic. Because phosphofructokinase (PFK) catalyzes the ATP-dependent phosphorylation to convert fructose-6-phosphate into fructose 1,6-bisphosphate and ADP, it is one of the key regulatory and rate limiting steps of glycolysis. PFK is able to regulate glycolysis through allosteric inhibition, and in this way, the cell can increase or decrease the rate of glycolysis in response to the cell's energy requirements. For example, a high ratio of ATP to ADP will inhibit PFK and glycolysis. The key difference between the regulation of PFK in eukaryotes and prokaryotes is that in eukaryotes PFK is activated by fructose 2,6-bisphosphate. The purpose of fructose 2,6-bisphosphate is to supersede ATP inhibition, thus allowing eukaryotes to have greater sensitivity to regulation by hormones like glucagon and insulin.

β-D-fructose 6-phosphate	Phosphofructokinase 1	β-D-fructose 1,6-bisphosphate
	ATP | ADP	
	P$_i$ | H$_2$O	
	Fructose bisphosphatase	

Structure

Mammalian PFK1 is a 340kd tetramer composed of different combinations of three types of subunits: muscle (M), liver (L), and platelet (P). The composition of the PFK1 tetramer differs according to the tissue type it is present in. For example, mature muscle expresses only the M isozyme, therefore, the muscle PFK1 is composed solely of homotetramers of M4. The liver and kidneys express predominantly the L isoform. In erythrocytes, both M and L subunits randomly tetramerize to form M4, L4 and the three hybrid forms of the enzyme (ML3, M2L2, M3L). As a result, the kinetic and regulatory properties of the various isoenzymes pools are dependent on subunit composition. Tissue-specific changes in PFK activity and isoenzymic content contribute significantly to the diversities of glycolytic and gluconeogenic rates which have been observed for different tissues.

PFK1 is an allosteric enzyme and has a structure similar to that of hemoglobin in so far as it is a dimer of a dimer. One half of each dimer contains the ATP binding site whereas the other half the substrate (fructose-6-phosphate or (F6P)) binding site as well as a separate allosteric binding site.

Each subunit of the tetramer is 319 amino acids and consists of two domain, one that binds the substrate ATP, and the other that binds fructose-6-phosphate. Each domain is a b barrel, and has cylindrical b sheet surrounded by alpha helices.

On the opposite side of the each subunit from each active site is the allosteric site, at the interface between subunits in the dimer. ATP and AMP compete for this site. The N-terminal domain has a catalytic role binding the ATP, and the C-terminal has a regulatory role

Mechanism

PFK1 is an allosteric enzyme whose activity can be described using the symmetry model of allosterism whereby there is a concerted transition from an enzymatically inactive T-state to the active R-state. F6P binds with a high affinity to the R state but not the T state enzyme. For every molecule of F6P that binds to PFK1, the enzyme progressively shifts from T state to the R state. Thus a graph plotting PFK1 activity against increasing F6P concentrations would adopt the sigmoidal curve shape traditionally associated with allosteric enzymes.

PFK1 belongs to the family of phosphotransferases and it catalyzes the transfer of γ-phosphate from ATP to fructose-6-phosphate. The PFK1 active site comprises both the ATP-Mg2+ and the F6P binding sites. Some proposed residues involved with substrate binding in *E. coli* PFK1 include Asp127 and Arg171. In B. stearothermophilus PFK1, the positively charged side chain of Arg162 residue forms a hydrogen-bonded salt bridge with the negatively charged phosphate group of F6P, an interaction which stabilizes the R state relative to the T state and is partly responsible for the homotropic effect of F6P binding. In the T state, enzyme conformation shifts slightly such that the space previously taken up by the Arg162 is replaced with Glu161. This swap in positions between adjacent amino acid residues inhibits the ability of F6P to bind the enzyme.

Allosteric activators such as AMP and ADP bind to the allosteric site as to facilitate the formation of the R state by inducing structural changes in the enzyme. Similarly, inhibitors such as ATP and PEP bind to the same allosteric site and facilitate the formation of the T state, thereby inhibiting enzyme activity.

The hydroxyl oxygen of carbon 1 does a nucleophilic attack on the beta phosphate of ATP. These electrons are pushed to the anhydride oxygen between the beta and gamma phosphates of ATP.

Mechanism of phosphofructokinase 1

Regulation

PFK1 is the most important control site in the mammalian glycolytic pathway. This step is subject to extensive regulation since it is not only highly exergonic under physiological conditions, but also because it is a committed step – the first irreversible reaction unique to the glycolytic pathway. This leads to a precise control of glucose and the other monosaccharides galactose and fructose going down the glycolytic pathway. Before this enzyme's reaction, glucose-6-phosphate can potentially travel down the pentose phosphate pathway, or be converted to glucose-1-phosphate for glycogenesis.

PFK1 is allosterically inhibited by high levels of ATP but AMP reverses the inhibitory action of ATP. Therefore, the activity of the enzyme increases when the cellular ATP/AMP ratio is lowered. Glycolysis is thus stimulated when energy charge falls. PFK1 has two sites with different affinities for ATP which is both a substrate and an inhibitor.

PFK1 is also inhibited by low pH levels which augment the inhibitory effect of ATP. The pH falls when muscle is functioning anaerobically and producing excessive quantities of lactic acid (although lactic acid is not itself the cause of the decrease in pH). This inhibitory effect serves to protect the muscle from damage that would result from the accumulation of too much acid.

Finally, PFK1 is allosterically inhibited by PEP, citrate, and ATP. Phosphoenolpyruvic acid is a product further downstream the glycolytic pathway. Although citrate does build up when the Krebs Cycle enzymes approach their maximum velocity, it is questionable whether citrate accumulates to a sufficient concentration to inhibit PFK-1 under normal physiological conditions. ATP concentration build up indicates an excess of energy and does have an allosteric modulation site on PFK1 where it decreases the affinity of PFK1 for its substrate.

PFK1 is allosterically activated by a high concentration of AMP, but the most potent activator is fructose 2,6-bisphosphate, which is also produced from fructose-6-phosphate by PFK2. Hence, an abundance of F6P results in a higher concentration of fructose 2,6-bisphosphate (F-2,6-BP). The binding of F-2,6-BP increases the affinity of PFK1 for F6P and diminishes the inhibitory effect of ATP. This is an example of feedforward stimulation as glycolysis is accelerated when glucose is abundant.

PFK is inhibited by glucagon through repression of synthesis. Glucagon activates protein kinase A which, in turn, shuts off the kinase activity of PFK2. This reverses any synthesis of F-2,6-BP from F6P and thus inhibits PFK1 activity.

The precise regulation of PFK1 prevents glycolysis and gluconeogenesis from occurring simultaneously. However, there is substrate cycling between F6P and F-1,6-BP. Fructose-1,6-bisphosphatase (FBPase) catalyzes the hydrolysis of F-1,6-BP back to F6P, the reverse reaction catalyzed by PFK1. There is a small amount of FBPase activity during glycolysis and some PFK1 activity during gluconeogenesis. This cycle allows for the amplification of metabolic signals as well as the generation of heat by ATP hydrolysis.

Serotonin (5-HT) increases PFK by binding to the 5-HT(2A) receptor, causing the tyrosine residue of PFK to be phosphorylated via phospholipase C. This in turn redistributes PFK within the skeletal muscle cells. Because PFK regulates glycolytic flux, serotonin plays a regulatory role in glycolysis

Clinical Significance

A genetic mutation in the PFKM gene results in Tarui's disease, which is a glycogen storage disease where the ability of certain cell types to utilize carbohydrates as a source of energy is impaired.

Tarui disease is a glycogen storage disease with symptoms including muscle weakness (myopathy) and exercise induced cramping and spasms, myoglobinuria (presence of myoglobin in urine, indicating muscle destruction) and compensated hemolysis. ATP is a natural allosteric inhibitor of PFK, in order to prevent unnecessary production of ATP through glycolysis. However, a mutation in Asp(543)Ala can result in ATP having a stronger inhibitory effect (due to increased binding to PFK's inhibitory allosteric binding site).

Phosphofructokinase mutation and cancer: In order for cancer cells to meet their energy requirements due to their rapid cell growth and division, they survive more effectively when they have a hyperactive phosphofructokinase 1 enzyme. When cancer cells grow and divide quickly, they initially do not have as much blood supply, and can thus have hypoxia (oxygen deprivation), and this triggers O-GlcNAcylation at serine 529 of PFK, giving a selective growth advantage to cancer cells.

Herpes simplex type 1 and phosphofructokinase: Some viruses, including HIV, HCMV, Mayaro, and HCMV affect cellular metabolic pathways such as glycolysis by a MOI-dependent increase in the activity of PFK. The mechanism that Herpes increases PFK activity is by phosphorylating the enzyme at the serine residues. The HSV-1 induced glycolysis increases ATP content, which is critical for the virus's replication.

Pentose Phosphate Pathway

In biochemistry, the pentose phosphate pathway (also called the phosphogluconate pathway and thehexose monophosphate shunt) is a metabolic pathway parallel to glycolysis that generates NADPH andpentoses (5-carbon sugars) as well as Ribose 5-phosphate, a precursor for the synthesis of nucleotides. While it does involve oxidation of glucose, its primary role is anabolic rather than catabolic.

There are two distinct phases in the pathway. The first is the oxidative phase, in which NADPH is generated, and the second is the non-oxidative synthesis of 5-carbon sugars. For most organisms, the pentose phosphate pathway takes place in the cytosol; in plants, most steps take place in plastids.

Similar to glycolysis, the pentose phosphate pathway appears to have a very ancient evolutionary origin. The reactions of this pathway are mostly enzyme-catalyzed in modern cells, however, they also occur non-enzymatically under conditions that replicate those of the Archean ocean, and are catalyzed by metal ions, particularly ferrous ions (Fe(II)). This suggests that the origins of the pathway could date back to the prebiotic world.

Outcome

The primary results of the pathway are:

- The generation of reducing equivalents, in the form of NADPH, used in reductive biosynthesis reactions within cells (e.g. fatty acid synthesis).

- Production of ribose 5-phosphate (R5P), used in the synthesis of nucleotides and nucleic acids.

- Production of erythrose 4-phosphate (E4P) used in the synthesis of aromatic amino acids.

Aromatic amino acids, in turn, are precursors for many biosynthetic pathways, including the lignin in wood.

Dietary pentose sugars derived from the digestion of nucleic acids may be metabolized through the pentose phosphate pathway, and the carbon skeletons of dietary carbohydrates may be converted into glycolytic/gluconeogenic intermediates.

In mammals, the PPP occurs exclusively in the cytoplasm, and is found to be most active in the liver, mammary gland and adrenal cortex in the human. The PPP is one of the three main ways the body creates molecules with reducing power, accounting for approximately 60% of NADPH production in humans.

One of the uses of NADPH in the cell is to prevent oxidative stress. It reduces glutathione via glutathione reductase, which converts reactive H_2O_2 into H_2O by glutathione peroxidase. If absent, the H_2O_2 would be converted to hydroxyl free radicals by Fenton chemistry, which can attack the cell. Erythrocytes, for example, generate a large amount of NADPH through the pentose phosphate pathway to use in the reduction of glutathione.

Hydrogen peroxide is also generated for phagocytes in a process often referred to as a respiratory burst.

Phases

Oxidative Phase

In this phase, two molecules of NADP$^+$ are reduced to NADPH, utilizing the energy from the conversion of glucose-6-phosphate into ribulose 5-phosphate.

Oxidative phase of pentose phosphate pathway.
Glucose-6-phosphate (**1**), 6-phosphoglucono-δ-
lactone (**2**), 6-phosphogluconate (**3**), ribulose 5-phosphate (**4**)

The entire set of reactions can be summarized as follows:

Reactants	Products	Enzyme	Description
Glucose 6-phosphate+ NADP+	→ 6-phosphoglucono-δ-lactone + **NADPH**	glucose 6-phosphate dehydrogenase	Dehydrogenation. The hydroxyl on carbon 1 of glucose 6-phosphate turns into a carbonyl, generating a lactone, and, in the process, NADPH is generated.

6-phosphoglucono-δ-lactone + H_2O	→ 6-phosphogluconate + H^+	6-phosphogluconolactonase	Hydrolysis
6-phosphogluconate+ NADP$^+$	→ ribulose 5-phosphate +**NADPH** + CO_2	6-phosphogluconate dehydrogenase	Oxidative decarboxylation. NADP$^+$ is the electron acceptor, generating another molecule ofNADPH, a CO_2, and ribulose 5-phosphate.

The overall reaction for this process is:

Glucose 6-phosphate + 2 NADP$^+$ + H_2O → ribulose 5-phosphate + 2 NADPH + 2 H$^+$ + CO_2

Non-Oxidative Phase

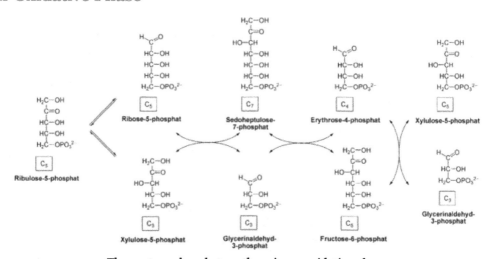

The pentose phosphate pathway's nonoxidative phase

Reactants	Products	Enzymes
ribulose 5-phosphate	→ ribose 5-phosphate	Ribulose 5-Phosphate Isomerase
ribulose 5-phosphate	→ xylulose 5-phosphate	Ribulose 5-Phosphate 3-Epimerase
xylulose 5-phosphate + ribose 5-phosphate	→ glyceraldehyde 3-phosphate + sedoheptulose 7-phosphate	transketolase
sedoheptulose 7-phosphate + glyceraldehyde 3-phosphate	→ erythrose 4-phosphate + fructose 6-phosphate	transaldolase
xylulose 5-phosphate + erythrose 4-phosphate	→ glyceraldehyde 3-phosphate + fructose 6-phosphate	transketolase

Net reaction: 3 ribulose-5-phosphate → 1 ribose-5-phosphate + 2 xylulose-5-phosphate $\rightarrow$ 2 fructose-6-phosphate + glyceraldehyde-3-phosphate

Regulation

Glucose-6-phosphate dehydrogenase is the rate-controlling enzyme of this pathway. It is allosterically stimulated by NADP$^+$ and strongly inhibited by NADPH. The ratio of NADPH:NADP$^+$ is normally about 100:1 in liver cytosol. This makes the cytosol a highly-reducing environment. An NADPH-utilizing pathway forms NADP$^+$, which stimulates Glucose-6-phosphate dehydrogenase to produce more NADPH. This step is also inhibited by acetyl CoA.

G6PD activity is also post-translationally regulated by cytoplasmic deacetylase SIRT2. SIRT2-mediated deacetylation and activation of G6PD stimulates oxidative branch of PPP to supply cytosolic NADPH to counteract oxidative damage or support *de novo* lipogenesis.

Erythrocytes and The Pentose Phosphate Pathway

Several deficiencies in the level of activity (not function) of glucose-6-phosphate dehydrogenase have been observed to be associated with resistance to the malarial parasite*Plasmodium falciparum* among individuals of Mediterranean and African descent. The basis for this resistance may be a weakening of the red cell membrane (the erythrocyte is the host cell for the parasite) such that it cannot sustain the parasitic life cycle long enough for productive growth.

Glucose-6-Phosphate Dehydrogenase

Glucose-6-phosphate dehydrogenase (G6PD or G6PDH) (EC 1.1.1.49) is a cytosolic enzyme that catalyzes the chemical reaction

D-glucose 6-phosphate + NADP$^+$ 6-phospho-D-glucono-1,5-lactone + NADPH + H$^+$

This enzyme participates in the pentose phosphate pathway, a metabolic pathway that supplies reducing energy to cells (such as erythrocytes) by maintaining the level of the co-enzyme nicotinamide adenine dinucleotide phosphate (NADPH). The NADPH in turn maintains the level of glutathione in these cells that helps protect the red blood cells against oxidative damage from compounds like hydrogen peroxide. Of greater quantitative importance is the production of NADPH for tissues actively engaged in biosynthesis of fatty acids and/or isoprenoids, such as the liver, mammary glands, adipose tissue, and the adrenal glands. G6PD reduces NADP$^+$ to NADPH while oxidizing glucose-6-phosphate.

Clinically, an X-linked genetic deficiency of G6PD predisposes a person to non-immune hemolytic anemia .

Species Distribution

G6PD is widely distributed in many species from bacteria to humans. Multiple sequence alignment of over 100 known G6PDs from different organisms reveal sequence identity ranging from 30% to 94%. Human G6PD has over 30% identity in amino acid sequence to G6DP sequences from other species. Humans also have two isoforms of a single gene coding for G6PD. Moreover, 150 different human G6PD mutants have been documented. These mutations are mainly missense mutations that result in amino acid substitutions, and while some of them result in G6PD deficiency, others do not seem to result in any noticeable functional differences. Some scientists have proposed that some of the genetic variation in human G6PD resulted from generations of adaptation to malarial infection.

Other species experience a variation in G6PD as well. In higher plants, several isoforms of G6PDH have been reported, which are localized in the cytosol, the plastidic stroma, and peroxisomes. A modified F_{420}-dependent (as opposed to $NADP^+$-dependent) G6PD is found in *Mycobacterium tuberculosis*, and is of interest for treating tuberculosis. The bacterial G6PD found in *Leuconostoc mesenteroides* was shown to be reactive toward 4-Hydroxynonenal, in addition to G6P.

Enzyme Structure

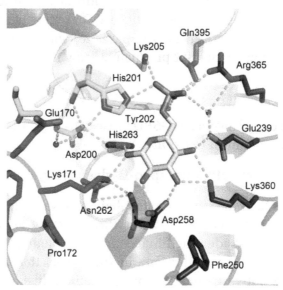

Substrate binding site of G6PD bound to G6P (shown in cream), from 2BHL. Phosphorus is shown in orange. Oxygen atoms of crystallographic waters are shown as red spheres. The conserved 9-peptide sequence of G6PD, and the partially conserved 5-residue sequence of G6PD are shown in cyan and magenta respectively. All other amino acids from G6PD are shown in black. Hydrogen bonding and electrostatic interactions are shown by green dashed lines. All green dashes represent distances of less than 3.7 Å.

G6PD is generally found as a dimer of two identical monomers. Depending on conditions, such as pH, these dimers can themselves dimerize to form tetramers.

Each monomer in the complex has a substrate binding site that binds to G6P, and a catalytic coenzyme binding site that binds to NADP⁺/NADPH using the Rossman fold. For some higher organisms, such as humans, G6PD contains an additional NADP⁺ binding site, called the NADP⁺ structural site, that does not seem to participate directly in the reaction catalyzed by G6PD. The evolutionary purpose of the NADP⁺ structural site is unknown. As for size, each monomers is approximately 500 amino acids long (514 amino acids for humans).

Functional and structural conservation between human G6PD and *Leuconostoc mesenteroides* G6PD points to 3 widely conserved regions on the enzyme: a 9 residue peptide in the substrate binding site, RIDHYLGKE (residues 198-206 on human G6PD), a nucleotide-binding fingerprint, GxxGDLA (residues 38-44 on human G6PD), and a partially conserved sequence EKPxG near the substrate binding site (residues 170-174 on human G6PD), where we have use "x" to denote a variable amino acid. The crystal structure of G6PD reveals an extensive network of electrostatic interactions and hydrogen bonding involving G6P, 3 water molecules, 3 lysines, 1 arginine, 2 histidines, 2 glutamic acids, and other polar amino acids.

The proline at position 172 is thought to play a crucial role in positioning Lys171 correctly with respect to the substrate, G6P. In the two crystal structures of normal normal human G6P, Pro172 is seen exclusively in the cis confirmation, while in the crystal structure of one disease causing mutant (variant Canton R459L), Pro172 is seen almost exclusively in the trans confirmation.

With access to crystal structures, some scientists have tried to model the structures of other mutants. For example, in German ancestry, where enzymopathy due to G6PD deficiency is rare, mutation sites on G6PD have been shown to lie near the NADP⁺ binding site, the G6P binding site, and near the interface between the two monomers. Thus, mutations in these critical areas are possible without completely disrupting the function of G6PD. In fact, it has been shown that most disease causing mutations of G6PD occur near the NADP⁺ structural site.

NADP⁺ Structural Site

The NADP⁺ structural site is located greater than 20Å away from the substrate binding site and the catalytic coenzyme NADP⁺ binding site. Its purpose in the enzyme catalyzed reaction has been unclear for many years. For some time, it was thought that NADP⁺ binding to the structural site was necessary for dimerization of the enzyme monomers. However, this was shown to be incorrect. On the other hand, it was shown that the presence of NADP⁺ at the structural site promotes the dimerization of dimers to form enzyme tetramers. It was also thought that the tetramer state was necessary for catalytic activity; however, this too was shown to be false. Interestingly, the NADP⁺ structural site is quite different from the NADP⁺ catalytic coenzyme binding site, and does contain the nucleotide-binding fingerprint.

The structural site bound to NADP$^+$ possesses favorable interactions that keep it tightly bound. In particular, there is a strong network of hydrogen bonding with electrostatic charges being diffused across multiple atoms through hydrogen bonding with 4 water molecules. Moreover, there is an extremely strong set of hydrophobic stacking interactions that result in overlapping π systems.

NADP$^+$ structural site of G6PD. NADP$^+$ is shown in cream. Phosphorus is shown in orange. The oxygen atoms of crystallographic water molecules are shown as red spheres. The conserved 9-peptide sequence of G6PD is show in cyan.

The structural site has been shown to be important for maintaining the long term stability of the enzyme. More than 40 severe class I mutations involve mutations near the structural site, thus affecting the long term stability of these enzymes in the body, ultimately resulting in G6PD deficiency. For example, two severe class I mutations, G488S and G488V, drastically increase the dissociation constant between NADP$^+$ and the structural site by a factor of 7 to 13. With the proximity of residue 488 to Arg487, it is thought that a mutation at position 488 could affect the positioning of Arg487 relative to NADP$^+$, and thus disrupt binding.

Regulation

G6PD converts G6P into 6-phosphoglucono-δ-lactone and is the rate-limiting enzyme of the *pentose phosphate pathway*. Thus, regulation of G6PD has downstream consequences for the activity of the rest of the *pentose phosphate pathway*.

Glucose-6-phosphate dehydrogenase is stimulated by its substrate G6P. The usual ratio of NADPH/NADP$^+$ in the cytosol of tissues engaged in biosyntheses is about 100/1. Increased utilization of NADPH for fatty acid biosynthesis will dramatically increase the level of NADP$^+$, thus stimulating G6PD to produce more NADPH.

G6PD is negatively regulated by acetylation on lysine 403 (Lys403), an evolutionarily conserved residue. The K403 acetylated G6PD is incapable of forming active dimers and displays a complete loss of activity. Mechanistically, acetylating Lys304 sterically hinders the NADP$^+$ from entering the NADP$^+$ structural site, which reduces the stability of the enzyme. Cells sense extracellular oxidative stimuli to decrease G6PD acetylation in a SIRT2-dependent manner. The SIRT2-mediated deacetylation and activation of G6PD stimulates pentose phosphate pathway to supply cytosolic NADPH to counteract oxidative damage and protect mouse erythrocytes.

Regulation can also occur through genetic pathways. The isoform, G6PDH, is regulated by transcription and posttranscription factors. Moreover, G6PD is one of a number of glycolytic enzymes activated by the transcription factor Hypoxia-inducible factor 1 (HIF1).

Clinical Significance

G6PD is remarkable for its genetic diversity. Many variants of G6PD, mostly produced from missense mutations, have been described with wide ranging levels of enzyme activity and associated clinical symptoms. Two transcript variants encoding different isoforms have been found for this gene.

Glucose-6-phosphate dehydrogenase deficiency is very common worldwide, and causes acute hemolytic anemia in the presence of simple infection, ingestion of fava beans, or reaction with certain medicines, antibiotics, antipyretics, and antimalarials.

Cell growth and proliferation are affected by G6PD. G6PD inhibitors are under investigation to treat cancers and other conditions. *In vitro* cell proliferation assay indicates that G6PD inhibitors, DHEA (dehydroepiandrosterone) and ANAD (6-aminonicotinamide), effectively decrease the growth of AML cell lines. G6PD is hypomethylated at K403 in acute myeloid leukemia, SIRT2 activates G6PD to enhance NADPH production and promote leukemia cell proliferation.

Fatty Acid Synthesis

Fatty acid synthesis is the creation of fatty acids from acetyl-CoA and NADPH with malonyl-CoA, through the action of enzymes called fatty acid synthases. This process takes place in the cytoplasm of the cell. Most of the acetyl-CoA which is converted into fatty acids is derived from carbohydrates via the glycolytic pathway. The glycolytic pathway also provides the glycerol with which three fatty acids can combine (by means of ester bonds) to form triglycerides (also known as "triacylglycerols", "neutral fats" - to distinguish them from fatty "acids" - or simply as "fat"), the final product of the lipogenic process. When only two fatty acids combine with glycerol and the third alcohol group is phosphorylated with a group such as phosphatidylcholine, a phospholipid is formed. Phospholipids form the bulk of the lipid bilayers that make up cell membranes and surround the organelles within the cells (e.g. the cell nucleus, mitochondria, endoplasmic reticulum, Golgi apparatus etc.)

Straight-Chain Fatty Acids

Straight-chain fatty acids occur in two types: saturated and unsaturated.

Saturated Straight-Chain Fatty Acids

The diagrams presented show how fatty acids are synthesized in microorganisms and list the enzymes found in Escherichia coli. These reactions are performed by fatty acid synthase II (FASII), which in general contain multiple enzymes that act as one complex. FASII is present in prokaryotes, plants, fungi, and parasites, as well as in mitochondria.

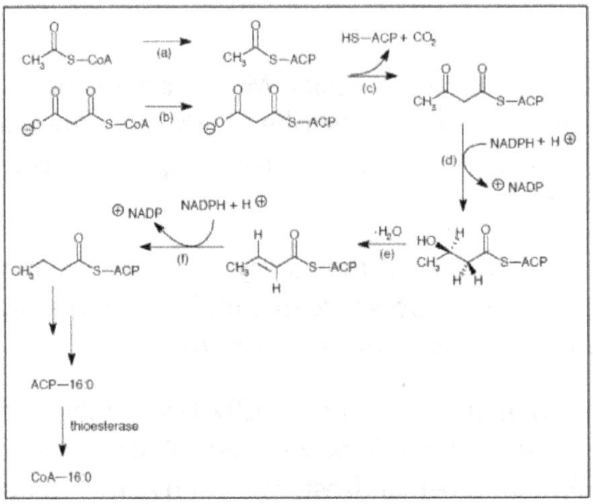

Synthesis of saturated fatty acids via Fatty Acid Synthase II in E. coli

Much like β-oxidation, straight-chain fatty acid synthesis occurs via the six recurring reactions shown below, until the 16-carbon palmitic acid is produced.

In animals, as well as some fungi such as yeast, these same reactions occur on fatty acid synthase I (FASI), a large dimeric protein that has all of the enzymatic activities required to create a fatty acid. FASI is less efficient than FASII; however, it allows for the formation of more molecules, including "medium-chain" fatty acids via early chain termination.

Once a 16:0 carbon fatty acid has been formed, it can undergo a number of modifications, resulting in desaturation and/or elongation. Elongation, starting with stearate (18:0), is performed mainly in the ER by several membrane-bound enzymes. The enzymatic steps involved in the elongation process are principally the same as those carried out by FAS, but the four principal successive steps of the elongation are performed by individual proteins, which may be physically associated.

Step	Enzyme	Reaction	Description
(a)	Acetyl CoA:ACP transacylase	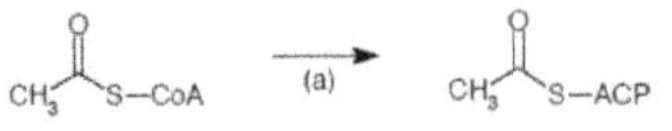	Activates acetyl CoA for reaction with malonyl-ACP
(b)	Malonyl CoA:ACP transacylase	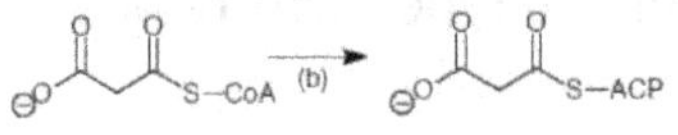	Activates malonyl CoA for reaction with acetyl-ACP
(c)	3-ketoacyl-ACP synthase	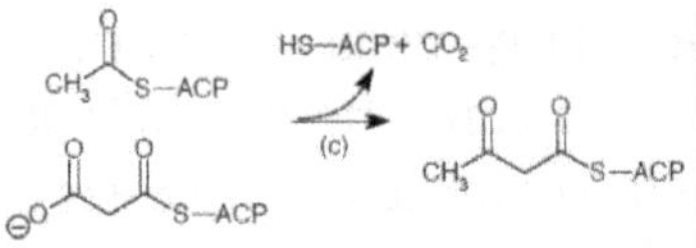	Reacts priming acetyl-ACP with chain-extending malonyl-ACP.

(d)	3-ketoacyl-ACP reductase		Reduces the carbon 3 ketone to a hydroxyl group
(e)	3-Hydroxyacyl ACP dehydrase		Removes water
(f)	Enoyl-ACP reductase		Reduces the C2-C3 double bond.

Abbreviations: ACP – Acyl carrier protein, CoA – Coenzyme A, NADP – Nicotinamide adenine dinucleotide phosphate.

Note that during fatty synthesis the reducing agent is NADPH, whereas NAD is the oxidizing agent in beta-oxidation (the breakdown of fatty acids to acetyl-CoA). This difference exemplifies a general principle that NADPH is consumed during biosynthetic reactions, whereas NADH is generated in energy-yielding reactions. (Thus NADPH is also required for the synthesis of cholesterol from acetyl-CoA; while NADH is generated during glycolysis.) The source of the NADPH is two-fold. When malate is oxidatively decarboxylated by "NADP⁺-linked malic enzyme" pyruvate, CO_2 and NADPH are formed. NADPH is also formed by the pentose phosphate pathway which converts glucose into ribose, which can be used in synthesis of nucleotides and nucleic acids, or it can be catabolized to pyruvate.

Glycolytic End Products are Used in The Conversion of Carbohydrates into Fatty Acids

In humans, fatty acids are formed from carbohydrates predominantly in the liver and adipose tissue, as well as in the mammary glands during lactation.

The pyruvate produced by glycolysis is an important intermediary in the conversion of carbohydrates into fatty acids and cholesterol. This occurs via the conversion of pyruvate into acetyl-CoA in the mitochondrion. However, this acetyl CoA needs to be transported into cytosol where the synthesis of fatty acids and cholesterol occurs. This cannot occur directly. To obtain cytosolic acetyl-CoA, citrate (produced by the condensation of acetyl CoA with oxaloacetate) is removed from the citric acid cycle and carried across the inner mitochondrial membrane into the cytosol. There it is cleaved by ATP citrate lyase into acetyl-CoA and oxaloacetate. The oxaloacetate can be used for gluconeogenesis (in the liver), or it can be returned into mitochondrion as malate. The cytosolic acetyl-CoA is carboxylated by acetyl CoA carboxylase into malonyl CoA, the first committed step in the synthesis of fatty acids.

Animals cannot Resynthesize Carbohydrates from Fatty Acids

The main fuel stored in the bodies of animals is fat. The young adult human's fat stores average between about 10-20 kg, but varies greatly depending on age, gender, and individual disposition. By contrast the human body stores only about 400 g of glycogen, of which 300 g is locked inside the skeletal muscles and is unavailable to the body as a whole. The 100 g or so of glycogen stored in the liver is depleted within one day of starvation. Thereafter the glucose that is released into the blood by the liver for general use by the body tissues, has to be synthesized from the glucogenic amino acids and a few other gluconeogenic substrates, which do not include fatty acids.

Fatty acids are broken down to acetyl-CoA by means of beta oxidation inside the mitochondria, whereas fatty acids are synthesized from acetyl-CoA outside the mitochondrion, in the cytosol. The two pathways are distinct, not only in where they occur, but also in the reactions that occur, and the substrates that are used. The two pathways are mutually inhibitory, preventing the acetyl-CoA produced by beta-oxidation from entering the synthetic pathway via the acetyl-CoA carboxylase reaction. It can also not be converted to pyruvate as the pyruvate decarboxylation reaction is irreversible. Instead it condenses with oxaloacetate, to enter the citric acid cycle. During each turn of the cycle, two carbon atoms leave the cycle as CO_2 in the decarboxylation reactions catalyzed by isocitrate dehydrogenase and alpha-ketoglutarate dehydrogenase. Thus each turn of the citric acid cycle oxidizes an acetyl-CoA unit while regenerating the oxaloacetate molecule with which the acetyl-CoA had originally combined to form citric acid. The decarboxylation reactions occur before malate is formed in the cycle. This is the only substance that can be removed from the mitochondrion to enter the gluconeogenic pathway to form glucose or glycogen in the liver or any other tissue. There can therefore be no net conversion of fatty acids into glucose.

Only plants possess the enzymes to convert acetyl-CoA into oxaloacetate from which malate can be formed to ultimately be converted to glucose.

Regulation

Acetyl-CoA is formed into malonyl-CoA by acetyl-CoA carboxylase, at which point malonyl-CoA is destined to feed into the fatty acid synthesis pathway. Acetyl-CoA carboxylase is the point of regulation in saturated straight-chain fatty acid synthesis, and is subject to both phosphorylation and allosteric regulation. Regulation by phosphorylation occurs mostly in mammals, while allosteric regulation occurs in most organisms. Allosteric control occurs as feedback inhibition by palmitoyl-CoA and activation by citrate. When there are high levels of palmitoyl-CoA, the final product of saturated fatty acid synthesis, it allosterically inactivates acetyl-CoA carboxylase to prevent a build-up of fatty acids in cells. Citrate acts to activate acetyl-CoA carboxylase under high levels, because high levels indicate that there is enough acetyl-CoA to feed into the Krebs cycle and produce energy.

High plasma levels of insulin in the blood plasma (e.g. after meals) cause the dephosphorylation of acetyl-CoA carboxylase, thus promoting the formation of malonyl-CoA from acetyl-CoA, and consequently the conversion of carbohydrates into fatty acids, while epinephrine and glucagon (released into the blood during starvation and exercise) cause the phosphorylation of this enzyme, inhibiting lipogenesis in favor of fatty acid oxidation via beta-oxidation.

Anaerobic Desaturation

Synthesis of unsaturated fatty acids via anaerobic desaturation

Many bacteria use the anaerobic pathway for synthesizing unsaturated fatty acids. This pathway does not utilize oxygen and is dependent on enzymes to insert the double bond before elongation utilizing the normal fatty acid synthesis machinery. In *Escherichia coli*, this pathway is well understood.

- FabA is a β-hydroxydecanoyl-ACP dehydrase – it is specific for the 10-carbon saturated fatty acid synthesis intermediate (β-hydroxydecanoyl-ACP).

- FabA catalyzes the dehydration of β-hydroxydecanoyl-ACP, causing the release of water and insertion of the double bond between C7 and C8 counting from the methyl end. This creates the trans-2-decenoyl intermediate.

- Either the trans-2-decenoyl intermediate can be shunted to the normal saturated fatty acid synthesis pathway by FabB, where the double bond will be hydrolyzed and the final product will be a saturated fatty acid, or FabA will catalyze the isomerization into the cis-3-decenoyl intermediate.

- FabB is a β-ketoacyl-ACP synthase that elongates and channels intermediates into the mainstream fatty acid synthesis pathway. When FabB reacts with the cis-decenoyl intermediate, the final product after elongation will be an unsaturated fatty acid.

- The two main unsaturated fatty acids made are Palmitoleoyl-ACP (16:1ω7) and cis-vaccenoyl-ACP (18:1ω7).

Most bacteria that undergo anaerobic desaturation contain homologues of FabA and FabB. Clostridia are the main exception; they have a novel enzyme, yet to be identified, that catalyzes the formation of the cis double bond.

Regulation

This pathway undergoes transcriptional regulation by FadR and FabR. FadR is the more extensively studied protein and has been attributed bifunctional characteristics. It acts as an activator of *fabA* and *fabB* transcription and as a repressor for the β-oxidation regulon. In contrast, FabR acts as a repressor for the transcription of fabA and fabB.

Aerobic Desaturation

Aerobic desaturation is the most widespread pathway for the synthesis of unsaturated fatty acids. It is utilized in all eukaryotes and some prokaryotes. This pathway utilizes desaturases to synthesize unsaturated fatty acids from full-length saturated fatty acid substrates. All desaturases require oxygen and ultimately consume NADH even though desaturation is an oxidative process. Desaturases are specific for the double bond they induce in the substrate. In Bacillus subtilis, the desaturase, Δ^5-Des, is specific for inducing a cis-double bond at the Δ^5 position. *Saccharomyces cerevisiae* contains one desaturase, Ole1p, which induces the cis-double bond at Δ^9.

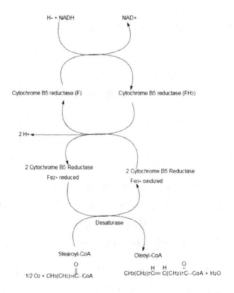

Synthesis of unsaturated fatty acids via aerobic desaturation

In mammals the aerobic desaturation is catalyzed by a complex of three mem-brane-bound enzymes (*NADH-cytochrome b_5 reductase, cytochrome b_5,* and a *desat-urase*). These enzymes allow molecular oxygen, O_2, to interact with the saturated fatty acyl-CoA chain, forming a double bond and two molecules of water, H_2O. Two electrons come from NADH + H^+ and two from the single bond in the fatty acid chain. These mammalian enzymes are, however, incapable of introducing double bonds at carbon atoms beyond C-9 in the fatty acid chain. Hence mammals cannot synthesize linoleate or linolenate (which have double bonds at the C-12 (= Δ^{12}), or the C-12 and C-15 (= Δ^{12} and Δ^{15}) positions, respectively, as well as at the Δ^9 position), nor the polyunsaturated, 20-carbon arachidonic acid that is derived from linoleate. These are all termed essential fatty acids, meaning that they are required by the organism, but can only be supplied via the diet. (Arachidonic acid is the precursor the prostaglandins which fulfill a wide variety of functions as local hormones.)

Regulation

In *B. subtilis*, this pathway is regulated by a two-component system: DesK and DesR. DesK is a membrane-associated kinase and DesR is a transcriptional regulator of the *des* gene. The regulation responds to temperature; when there is a drop in temperature, this gene is upregulated. Unsaturated fatty acids increase the fluidity of the membrane and stabilize it under lower temperatures. DesK is the sensor protein that, when there is a decrease in temperature, will autophosphorylate. DesK-P will transfer its phosphoryl group to DesR. Two DesR-P proteins will dimerize and bind to the DNA promoters of the *des* gene and recruit RNA polymerase to begin transcription.

Pseudomonas Aeruginosa

In general, both anaerobic and aerobic unsaturated fatty acid synthesis will not oc-cur within the same system, however *Pseudomonas aeruginosa* and *Vibrio* ABE-1 are exceptions. While *P. aeruginosa* undergoes primarily anaerobic desaturation, it also undergoes two aerobic pathways. One pathway utilizes a Δ^9-desaturase (DesA) that catalyzes a double bond formation in membrane lipids. Another pathway uses two pro-teins, DesC and DesB, together to act as a Δ^9-desaturase, which inserts a double bond into a saturated fatty acid-CoA molecule. This second pathway is regulated by repressor protein DesT. DesT is also a repressor of *fabAB* expression for anaerobic desaturation when in presence of exogenous unsaturated fatty acids. This functions to coordinate the expression of the two pathways within the organism.

Branched-Chain Fatty Acids

Branched-chain fatty acids are usually saturated and are found in two distinct families: the iso-series and anteiso-series. It has been found that Actinomycetales contain unique branch-chain fatty acid synthesis mechanisms, including that which forms tuberculosteric acid.

The branched-chain fatty acid synthesizing system uses α-keto acids as primers. This system is distinct from the branched-chain fatty acid synthetase that utilizes short-chain acyl-CoA esters as primers. α-Keto acid primers are derived from the transamination and decarboxylation of valine, leucine, and isoleucine to form 2-methylpropanyl-CoA, 3-methylbutyryl-CoA, and 2-Methylbutyryl-CoA, respectively. 2-Methylpropanyl-CoA primers derived from valine are elongated to produce even-numbered iso-series fatty acids such as 14-methyl-pentadecanoic (isopalmitic) acid, and 3-methylbutyryl-CoA primers from leucine may be used to form odd-numbered iso-series fatty acids such as 13-methyl-tetradecanoic acid. 2-Methylbutyryl-CoA primers from isoleucine are elongated to form anteiso-series fatty acids containing an odd number of carbon atoms such as 12-Methyl tetradecanoic acid. Decarboxylation of the primer precursors occurs through the branched-chain α-keto acid decarboxylase (BCKA) enzyme. Elongation of the fatty acid follows the same biosynthetic pathway in *Escherichia coli* used to produce straight-chain fatty acids where malonyl-CoA is used as a chain extender. The major end products are 12–17 carbon branched-chain fatty acids and their composition tends to be uniform and characteristic for many bacterial species.

Bcka Decarboxylase and Relative Activities of A-Keto Acid Substrates

The BCKA decarboxylase enzyme is composed of two subunits in a tetrameric structure (A_2B_2) and is essential for the synthesis of branched-chain fatty acids. It is responsible for the decarboxylation of α-keto acids formed by the transamination of valine, leucine, and isoleucine and produces the primers used for branched-chain fatty acid synthesis. The activity of this enzyme is much higher with branched-chain α-keto acid substrates than with straight-chain substrates, and in Bacillus species its specificity is highest for the isoleucine-derived α-keto-β-methylvaleric acid, followed by α-ketoisocaproate and α-ketoisovalerate. The enzyme's high affinity toward branched-chain α-keto acids allows it to function as the primer donating system for branched-chain fatty acid synthetase.

Substrate	BCKA activity	CO2 Produced (nmol/ min mg)	Km (μM)	Vmax (nmol/min mg)
$_L$-α-keto-β-methyl-valerate	100%	19.7	<1	17.8
α-Ketoisovalerate	63%	12.4	<1	13.3
α-Ketoisocaproate	38%	7.4	<1	5.6
Pyruvate	25%	4.9	51.1	15.2

Factors Affecting Chain Length and Pattern Distribution

α-Keto acid primers are used to produce branched-chain fatty acids that, in general, are between 12 and 17 carbons in length. The proportions of these branched-chain fatty acids tend to be uniform and consistent among a particular bacterial species but may be altered due to changes in malonyl-CoA concentration, temperature, or heat-stable factors (HSF) present. All of these factors may affect chain length, and HSFs have been demonstrated to alter the specificity of BCKA decarboxylase for a particular α-keto acid substrate, thus shifting the ratio of branched-chain fatty acids produced. An increase in malonyl-CoA concentration has been shown to result in a larger proportion of C17 fatty acids produced, up until the optimal concentration ($\approx 20\mu M$) of malonyl-CoA is reached. Decreased temperatures also tend to shift the fatty-acid distribution slightly toward C17 fatty-acids in *Bacillus* species.

Branch-Chain Fatty Acid Synthase

This system functions similarly to the branch-chain fatty acid synthesizing system, however it uses short-chain carboxylic acids as primers instead of alpha-keto acids. In general, this method is used by bacteria that do not have the ability to perform the branch-chain fatty acid system using alpha-keto primers. Typical short-chain primers include isovalerate, isobutyrate, and 2-methyl butyrate. In general, the acids needed for these primers are taken up from the environment; this is often seen in ruminal bacteria.

The overall reaction is:

Isobutyryl-CoA + 6 malonyl-CoA +12 NADPH + 12H$^+$ → Isopalmitic acid + 6 CO_2 12 NADP + 5 H_2O + 7 CoA

The difference between (straight-chain) fatty acid synthase and branch-chain fatty acid synthase is substrate specificity of the enzyme that catalyzes the reaction of acyl-CoA to acyl-ACP.

Omega-Alicyclic Fatty Acids

Omega-alicyclic fatty acids typically contain an omega-terminal propyl or butyryl cyclic group and are some of the major membrane fatty acids found in several species of bacteria. The fatty acid synthetase used to produce omega-alicyclic fatty acids is also used to produce membrane branched-chain fatty acids. In bacteria with membranes composed mainly of omega-alicyclic fatty acids, the supply of cyclic carboxylic acid-CoA esters is much greater than that of branched-chain primers. The synthesis of cyclic primers is not well understood but it has been suggested that mechanism involves the conversion of sugars to shikimic acid which is then converted to cyclohexylcarboxylic acid-CoA esters that serve as primers for omega-alicyclic fatty acid synthesis

11-cyclohexylundecanoic acid
Omega-alicyclic fatty acid

Tuberculostearic Acid Synthesis

Tuberculostearic acid ($_D$-10-Methylstearic acid) is a saturated fatty acid that is known to be produced by Mycobacterium spp. and two species of *Streptomyces*. It is formed from the precursor oleic acid (a monosaturated fatty acid). After oleic acid is esterified to a phospholipid, S-adenosyl-methionine donates a methyl group to the double bond of oleic acid. This methylation reaction forms the intermediate 10-methylene-octadecanoyal. Successive reduction of the residue, with NADPH as a cofactor, results in 10-methylstearic acid

Mechanism of the synthesis of Tuberculostearic acid

Fatty Acid Synthase

Fatty acid synthase (FAS) is an enzyme that in humans is encoded by the *FASN* gene.

Fatty acid synthase is a multi-enzyme protein that catalyzes fatty acid synthesis. It is not a single enzyme but a whole enzymatic system composed of two identical 272 kDa multifunctional polypeptides, in which substrates are handed from one functional domain to the next.

Its main function is to catalyze the synthesis of palmitate (C16:0, a long-chain saturated fatty acid) from acetyl-CoA and malonyl-CoA, in the presence of NADPH.

Metabolic Function

Fatty acids are aliphatic acids fundamental to energy production and storage, cellular structure and as intermediates in the biosynthesis of hormones and other biologically important molecules. They are synthesized by a series of decarboxylative Claisen condensation reactions from acetyl-CoA and malonyl-CoA. Following each round of elongation the beta keto group is reduced to the fully saturated carbon chain by the sequential action of a ketoreductase (KR), dehydratase (DH), and enoyl reductase (ER). The growing fatty acid chain is carried between these active sites while attached covalently to the phosphopantetheine prosthetic group of an acyl carrier protein (ACP), and is released by the action of a thioesterase (TE) upon reaching a carbon chain length of 16 (palmitidic acid).

Classes

There are two principal classes of fatty acid synthases.

- Type I systems utilise a single large, multifunctional polypeptide and are common to both mammals and fungi (although the structural arrangement of fungal and mammalian synthases differ). A Type I fatty acid synthase system is also found in the CMN group of bacteria (corynebacteria, mycobacteria, and nocardia). In these bacteria, the FAS I system produces palmititic acid, and cooperates with the FAS II system to produce a greater diversity of lipid products.

- Type II is found in archaea and bacteria, and is characterized by the use of discrete, monofunctional enzymes for fatty acid synthesis. Inhibitors of this pathway (FASII) are being investigated as possible antibiotics.

The mechanism of FAS I and FAS II elongation and reduction is the same, as the domains of the FAS II enzymes are largely homologous to their domain counterparts in FAS I multienzyme polypeptides. However, the differences in the organization of the enzymes - integrated in FAS I, discrete in FAS II - gives rise to many important biochemical differences.

The evolutionary history of fatty acid synthases are very much intertwined with that of polyketide synthases (PKS). Polyketide synthases use a similar mechanism and homologous domains to produce secondary metabolite lipids. Furthermore, polyketide synthases also exhibit a Type I and Type II organization. FAS I in animals is thought to have arisen through modification of PKS I in fungi, whereas FAS I in fungi and the CMN group of bacteria seem to have arisen separately through the fusion of FAS II genes.

Structure

Mammalian FAS consists of a homodimer of two identical protein subunits, in which three catalytic domains in the N-terminal section (-ketoacyl synthase (KS), malonyl/ acetyltransferase (MAT), and dehydrase (DH)), are separated by a core region of 600 residues from four C-terminal domains (enoyl reductase (ER), -ketoacyl reductase (KR), acyl carrier protein (ACP) and thioesterase (TE)).

The conventional model for organization of FAS is largely based on the observations that the bifunctional reagent 1,3-dibromo-propanone (DBP) is able to crosslink the active site cysteine thiol of the KS domain in one FAS monomer with the phosphopantetheine prosthetic group of the ACP domain in the other monomer. Complementation analysis of FAS dimers carrying different mutations on each monomer has established that the KS and MAT domains can cooperate with the ACP of either monomer. and a reinvestigation of the DBP crosslinking experiments revealed that the KS active site Cys161 thiol could be crosslinked to the ACP 4'-phosphopantetheine thiol of either monomer. In addition, it has been recently reported that a heterodimeric FAS containing only one competent monomer is capable of palmitate synthesis.

The above observations seemed incompatible with the classical 'head-to-tail' model for FAS organization, and an alternative model has been proposed, predicting that the KS and MAT domains of both monomers lie closer to the center of the FAS dimer, where they can access the ACP of either subunit.

A low resolution X-ray crystallography structure of both pig (homodimer) and yeast FAS (heterododecamer) along with a ~6 Å resolution electron cryo-microscopy (cryo-EM) yeast FAS structure have been solved.

Substrate Shuttling Mechanism

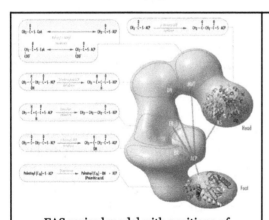

FAS revised model with positions of polypeptides, three catalytic domains and their corresponding reactions, visualization by Kosi Gramatikoff. Note that FAS is only active as a homodimer rather than the monomer pictured.

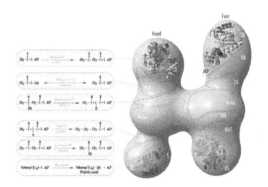

FAS 'head-to-tail' model with positions of polypeptides, three catalytic domains and their corresponding reactions, visualization by Kosi Gramatikoff.

The solved structures of yeast FAS and mammalian FAS show two distinct organization of highly conserved catalytic domains/enzymes in this multi-enzyme cellular machine. Yeast FAS has a highly efficient rigid barrel-like structure with 6 reaction chambers which synthesize fatty acids independently, while the mammalian FAS has an open flexible structure with only two reaction chambers. However, in both cases the conserved ACP acts as the mobile domain responsible for shuttling the intermediate fatty acid substrates to various catalytic sites. A first direct structural insight into this substrate shuttling mechanism was obtained by cryo-EM analysis, where ACP is observed bound to the various catalytic domains in the barrel-shaped yeast fatty acid synthase. The cryo-EM results suggest that the binding of ACP to various sites is asymmetric and stochastic, as also indicated by computer-simulation studies

Regulation

Metabolism and homeostasis of fatty acid synthase is transcriptionally regulated by Upstream Stimulatory Factors (USF1 and USF2) and sterol regulatory element binding protein-1c (SREBP-1c) in response to feeding/insulin in living animals.

Although liver X receptor (LXRs) modulate the expression of sterol regulatory element binding protein-1c (SREBP-1c) in feeding, regulation of FAS by SREBP-1c is USF-dependent.

Acylphloroglucinols isolated from the fern *Dryopteris crassirhizoma* show a fatty acid synthase inhibitory activity.

Clinical Significance

The gene that codes for FAS has been investigated as a possible oncogene. FAS is upregulated in breast cancers and as well as being an indicator of poor prognosis may also be worthwhile as a chemotherapeutic target. FAS may also be involved in the production of an endogenous ligand (biochemistry) for the nuclear receptor PPARalpha, the target of the fibrate drugs for hyperlipidemia, and is being investigated as a possible drug target for treating the metabolic syndrome.

In some cancer cell lines, this protein has been found to be fused with estrogen receptor alpha (ER-alpha), in which the N-terminus of FAS is fused in-frame with the C-terminus of ER-alpha.

An association with uterine leiomyomata has been reported.

References

- Pratt, Donald Voet, Judith G. Voet, Charlotte W. (2013). Fundamentals of Biochemistry: Life at the Molecular Level (4th ed.). Hoboken, NJ: Wiley. pp. 441–442. ISBN 978-0470-54784-7.

- Clarke, Jeremy M. Berg; John L. Tymoczko; Lubert Stryer. Web content by Neil D. (2002). Biochemistry (5. ed., 4. print. ed.). New York, NY [u.a.]: W. H. Freeman. pp. 578–579. ISBN 0716730510.

- Eichhorn, Peter H. Raven ; Ray F. Evert ; Susan E. (2011). Biology of plants (8. ed.). New York, NY: Freeman. pp. 100–106. ISBN 978-1-4292-1961-7.

- Pratt, Donald Voet, Judith G. Voet, Charlotte W. (2013). Fundamentals of biochemistry : life at the molecular level (4th ed.). Hoboken, NJ: Wiley. pp. 474–478. ISBN 978-0470-54784-7.

- Clarke, Jeremy M. Berg; John L. Tymoczko; Lubert Stryer. Web content by Neil D. (2002). Biochemistry (5. ed., 4. print. ed.). New York, NY [u.a.]: W. H. Freeman. p. 570. ISBN 0716730510.

- Berg, Jeremy M.; Tymoczko, John L.; Stryer, Lubert; Gatto, Gregory J. (2012). Biochemistry (7th ed.). New York: W.H. Freeman. pp. 480–482. ISBN 9781429229364.

- Clarke, Jeremy M. Berg; John L. Tymoczko; Lubert Stryer. Web content by Neil D. (2002). Biochemistry (5. ed., 4. print. ed.). New York, NY [u.a.]: W. H. Freeman. p. 570. ISBN 0-7167-3051-0.

- Pratt, Donald Voet, Judith G. Voet, Charlotte W. (2013). Fundamentals of biochemistry : life at the molecular level (4th ed.). Hoboken, NJ: Wiley. p. 572. ISBN 978-0470-54784-7.

- Cox, David L. Nelson, Michael M. (2008). Lehninger principles of biochemistry (5th ed.). New York: W.H. Freeman. pp. 577–578. ISBN 978-0-7167-7108-1.

- Krebs HA, Weitzman PD (1987). Krebs' citric acid cycle: half a century and still turning. London: Biochemical Society. p. 25. ISBN 0-904498-22-0.

- Jones RC, Buchanan BB, Gruissem W (2000). Biochemistry & molecular biology of plants (1st ed.). Rockville, Md: American Society of Plant Physiologists. ISBN 0-943088-39-9.

- Stryer L, Berg JM, Tymoczko JL (2002). "Section 18.6: The Regulation of Cellular Respiration Is Governed Primarily by the Need for ATP". Biochemistry. San Francisco: W.H. Freeman. ISBN 0-7167-4684-0.

6

Causes of Enzyme Deficiency

There are several causes behind enzyme depletion; this chapter studies the causes and the effects of enzyme depletion in detail. The chapter contains topics like chronic granulomatous disease, cortisone reductase deficiency, hypophosphatasia, myeloperoxidase deficiency, neutrophil-specific granule deficiency, phenylketonuria and pseudocholinesterase deficiency among others. The major components of enzyme deficiency are discussed in this chapter.

Chronic Granulomatous Disease

Chronic granulomatous disease (CGD) (also known as Bridges–Good syndrome, Chronic granulomatous disorder, and Quie syndrome) is a diverse group of hereditary diseases in which certain cells of the immune system have difficulty forming the reactive oxygen compounds (most importantly the superoxide radical due to defective phagocyte NADPH oxidase) used to kill certain ingested pathogens. This leads to the formation of granulomata in many organs. CGD affects about 1 in 200,000 people in the United States, with about 20 new cases diagnosed each year.

This condition was first discovered in 1950 in a series of 4 boys from Minnesota, and in 1957 was named "a fatal granulomatosus of childhood" in a publication describing their disease. The underlying cellular mechanism that causes chronic granulomatous disease was discovered in 1967, and research since that time has further elucidated the molecular mechanisms underlying the disease. Bernard Babior made key contributions in linking the defect of superoxide production of white blood cells, to the etiology of the disease. In 1986, the X-linked form of CGD was the first disease for which positional cloning was used to identify the underlying genetic mutation.

Classification

Chronic granulomatous disease is the name for a genetically heterogeneous group of immunodeficiencies. The core defect is a failure of phagocytic cells to kill organisms that they have engulfed because of defects in a system of enzymes that produce free radicals and other toxic small molecules. There are several types, including:

- X-linked chronic granulomatous disease]] (CGD)
- autosomal recessive cytochrome b-negative CGD

- autosomal recessive cytochrome b-positive CGD type I

- autosomal recessive cytochrome b-positive CGD type II

- atypical granulomatous disease

Symptoms

Classically, patients with chronic granulomatous disease will suffer from recurrent bouts of infection due to the decreased capacity of their immune system to fight off disease-causing organisms. The recurrent infections they acquire are specific and are, in decreasing order of frequency:

- pneumonia

- abscesses of the skin, tissues, and organs

- suppurative arthritis

- osteomyelitis

- bacteremia/fungemia

- superficial skin infections such as cellulitis or impetigo

Most people with CGD are diagnosed in childhood, usually before age 5. Early diagnosis is important since these people can be placed on antibiotics to ward off infections before they occur. Small groups of CGD patients may also be affected by McLeod syndrome because of the proximity of the two genes on the same X-chromosome.

Atypical Infections

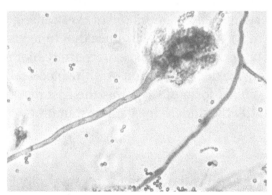

Microscopic image of the fungus, *Aspergillus fumigatus*, an organism that commonly causes disease in people with chronic granulomatous disease.

People with CGD are sometimes infected with organisms that usually do not cause disease in people with normal immune systems. Among the most common organisms that cause disease in CGD patients are:

- bacteria (particularly those that are catalase-positive)

  - *Staphylococcus aureus.*

  - *Serratia marcescens.*

  - *Listeria* species.

  - *E. coli.*

  - *Klebsiella* species.

  - *Pseudomonas cepacia, a.k.a. Burkholderia cepacia.*

  - *Nocardia.*

- fungi

  - *Aspergillus* species. Aspergillus has a propensity to cause infection in people with CGD and of the Aspergillus species, *Aspergillus fumigatus* seems to be most common in CGD.

  - *Candida* species.

Patients with CGD can usually resist infections of catalase-negative bacteria. Catalase is an enzyme that catalyzes the breakdown of hydrogen peroxide in many organisms. In organisms that lack catalase (catalase-negative), normal metabolic functions will cause an accumulation of hydrogen peroxide which the host's immune system can use to fight off the infection. In organisms that have catalase (catalase-positive), the enzyme breaks down any hydrogen peroxide that was produced through normal metabolism. Therefore, hydrogen peroxide will not accumulate, leaving the patient vulnerable to catalase-positive bacteria.

Genetics

Most cases of chronic granulomatous disease are transmitted as a mutation on the X chromosome and are thus called an "X-linked trait". The affected gene on the X chromosome codes for the gp91 protein p91-PHOX (*p* is the weight of the protein in kDa; the *g* means glycoprotein). CGD can also be transmitted in an autosomal recessive fashion (via CYBA and NCF1) and affects other PHOX proteins. The type of mutation that causes both types of CGD are varied and may be deletions, frame-shift, nonsense, and missense.

A low level of NADPH, the cofactor required for superoxide synthesis, can lead to CGD. This has been reported in women who are homozygous for the genetic defect causing glucose-6-phosphate dehydrogenase deficiency (G6PD), which is characterised by reduced NADPH levels.

Pathophysiology

Phagocytes (i.e., neutrophils and macrophages) require an enzyme to produce reactive oxygen species to destroy bacteria after they are ingested (phagocytosis), a process known as the respiratory burst. This enzyme is termed "phagocyte NADPH oxidase" (*PHOX*). This enzyme oxidizes NADPH and reduces molecular oxygen to produce superoxide anions, a reactive oxygen species. Superoxide is then disproportionated into peroxide and molecular oxygen by superoxide dismutase. Finally, peroxide is used by myeloperoxidase to oxidize chloride ions into hypochlorite (the active component of bleach), which is toxic to bacteria. Thus, NADPH oxidase is critical for phagocyte killing of bacteria through reactive oxygen species.

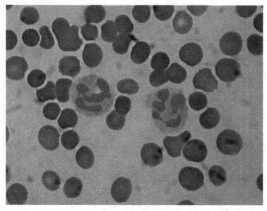

Two neutrophils among many red blood cells. Neutrophils are one type of cell affected by chronic granulomatous disease.

(Two other mechanisms are used by phagocytes to kill bacteria: nitric oxide and proteases, but the loss of ROS-mediated killing alone is sufficient to cause chronic granulomatous disease.)

Defects in one of the four essential subunits of phagocyte NADPH oxidase (PHOX) can all cause CGD of varying severity, dependent on the defect. There are over 410 known possible defects in the PHOX enzyme complex that can lead to chronic granulomatous disease.

Diagnosis

The nitroblue-tetrazolium (NBT) test is the original and most widely known test for chronic granulomatous disease. It is negative in CGD, meaning that it does not turn blue. The higher the blue score, the better the cell is at producing reactive oxygen species. This test depends upon the direct reduction of NBT to the insoluble blue compound formazan by NADPH oxidase; NADPH is oxidized in the same reaction. This test is simple to perform and gives rapid results, but only tells whether or not there is a problem with the PHOX enzymes, not how much they are affected. A similar test uses dihydrorhodamine (DHR) where whole blood is stained with DHR, incubated, and stimulated to produce superoxide radicals which oxidize DHR to rhodamin in cells with normal function. An advanced test called the cytochrome C reduction assay tells physicians how much superoxide a patient's phagocytes can produce. Once the diagnosis of CGD is established, a genetic analysis may be used to determine exactly which mutation is the underlying cause.

Treatment

Management of chronic granulomatous disease revolves around two goals: 1) diagnose the disease early so that antibiotic prophylaxis can be given to keep an infection from occurring, and 2) educate the patient about his or her condition so that prompt treatment can be given if an infection occurs.

Antibiotics

Physicians often prescribe the antibiotic trimethoprim-sulfamethoxazole to prevent bacterial infections. This drug also has the benefit of sparing the normal bacteria of the digestive tract. Fungal infection is commonly prevented with itraconazole, although a newer drug of the same type called voriconazole may be more effective. The use of this drug for this purpose is still under scientific investigation.

Immunomodulation

Interferon, in the form of interferon gamma-1b (Actimmune) is approved by the Food and Drug Administration for the prevention of infection in CGD. It has been shown to reduce infections in CGD patients by 70% and to decrease their severity. Although its exact mechanism is still not entirely understood, it has the ability to give CGD patients more immune function and therefore, greater ability to fight off infections. This therapy has been standard treatment for CGD for several years.

Hematopoietic Stem Cell Transplantation (Hsct)

Hematopoietic stem cell transplantation from a matched donor is curative although not without significant risk.

Prognosis

There are currently no studies detailing the long term outcome of chronic granulomatous disease with modern treatment. Without treatment, children often die in the first decade of life. The increased severity of X-linked CGD results in a decreased survival rate of patients, as 20% of X-linked patients die of CGD-related causes by the age of 10, whereas 20% of autosomal recessive patients die by the age of 35. Recent experience from centers specializing in the care of patients with CGD suggests that the current mortality has fallen to under 3% and 1% respectively. CGD was initially termed "fatal granulomatous disease of childhood" because patients rarely survived past their first decade in the time before routine use of prophylactic antimicrobial agents. The average patient now survives at least 40 years.

Epidemiology

CGD affects about 1 in 200,000 people in the United States, with about 20 new cases diagnosed each year.

Chronic granulomatous disease affects all people of all races, however, there is limited information on prevalence outside of the United States. One survey in Sweden reported an incidence of 1 in 220,000 people, while a larger review of studies in Europe suggested a lower rate: 1 in 250,000 people.

History

This condition was first described in 1954 by Janeway, who reported five cases of the disease in children. In 1957 it was further characterized as "a fatal granulomatosus of childhood". The underlying cellular mechanism that causes chronic granulomatous disease was discovered in 1967, and research since that time has further elucidated the molecular mechanisms underlying the disease. Use of antibiotic prophylaxis, surgical abscess drainage, and vaccination led to the term "fatal" being dropped from the name of the disease as children survived into adulthood.

Research

Gene therapy is currently being studied as a possible treatment for chronic granulomatous disease. CGD is well-suited for gene therapy since it is caused by a mutation in single gene which only affects one body system (the hematopoietic system). Viruses have been used to deliver a normal gp91 gene to rats with a mutation in this gene, and subsequently the phagocytes in these rats were able to produce oxygen radicals.

In 2006, two human patients with X-linked chronic granulomatous disease underwent gene therapy and blood cell precursor stem cell transplantation to their bone marrow. Both patients recovered from their CGD, clearing pre-existing infections and demonstrating increased oxidase activity in their neutrophils. However, long-term complications and efficacy of this therapy were unknown.

In 2012, a 16-year-old boy with CGD was treated at the Great Ormond Street Hospital, London with an experimental gene therapy which temporarily reversed the CGD and allowed him to overcome a life-threatening lung disease.

Cortisone Reductase Deficiency

Cortisone reductase deficiency is caused by dysregulation of the 11β-hydroxysteroid dehydrogenase type 1 enzyme (11β-HSD1), otherwise known as cortisone reductase, a bi-directional enzyme, which catalyzes the interconversion of cortisone to cortisol in the presence of NADH as a co-factor. If levels of NADH are low, the enzyme catalyses the reverse reaction, from cortisol to cortisone, using NAD+ as a co-factor. Cortisol is a glucocorticoid that plays a variety of roles in many different biochemical pathways, including, but not limited to: gluconeogenesis, suppressing immune system responses and carbohydrate metabolism. One of the symptoms of cortisone reductase deficiency is hyperandrogenism, resulting from activation of the Hypothalamic–pituitary–adrenal axis. The deficiency has been known to exhibit symptoms of other disorders such as Polycystic Ovary Syndrome in women. Cortisone Reductase Deficiency alone has been reported in fewer than ten cases in total, all but one case were women. Elevated activity of 11β-HSD1 can lead to obesity or Type II Diabetes, because of the role of cortisol in carbohydrate metabolism and gluconeogenesis.

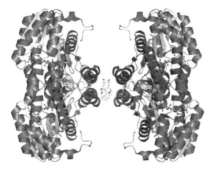

11β-hydroxysteroid dehydrogenase type 1

Cause

Pathophysiology

In a healthy body, blood cortisone and cortisol levels are roughly equimolar. Cortisone reductase deficiency leads to an elevated level of inert cortisone to active cortisol in adipose tissue. Cortisone reductase deficiency is caused by dysregulation of the 11β-hydroxysteroid dehydrogenase type 1 enzyme, otherwise known as cortisone reductase. The 11β-HSD1 enzyme is responsible for catalyzing the interconversion of circulating cortisone to cortisol, using NADH as a co-factor. The oxidative or reductive capacity of the enzyme is regulated by NADH produced by hexose-6-phosphate dehydrogenase (H6PD). H6PD is distinct from its isozyme, glucose-6-phosphate dehydrogenase (G6PDH) in that G6PDH is a cytolytic enzyme and draws from a separate pool of NAD+. H6PD is also capable of catalyzing the oxidation of several phosphorylated hexoses, while G6PDH shows affinity for glucose, specifically. The enzyme cortisone reductase exists in a tightly controlled reaction space, facing the lumen of the endoplasmic reticulum of cells in the liver and lungs. NADH produced by hexose-6-phosphate is delivered directly to the catalytic site of cortisone reductase. If NADH production is limited, then cortisone reductase is also capable of catalysing the reverse reaction taking circulating cortisol and reducing it to cortisone. Dysregulation of hexose-6-phosphate dehydrogenase occurs as a result of gene mutation. Cortisol is important in signalling inhibition of adrenocorticotropic hormone release from the pituitary. Reduced cortisol in circulation activates the H-P-A Axis to produce and release more cortisol, and therefore androgen.

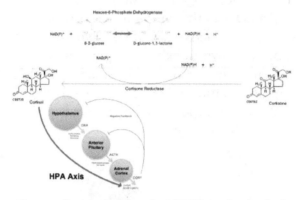

An overview of how cortisone reductase is driven by NADH production by hexose-6-phosphate and how it affects the HPA Axis in a healthy body.

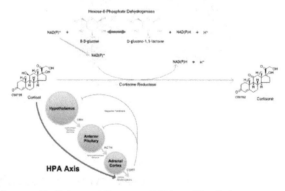

Cortisone Reductase Deficiency effects on HPA and body in presence of deficient H6PD

Effect on Hpa Axis

The Hypothalamic-Pituitary-Adrenal axis relies on blood levels of cortisol to act as negative feedback. Low levels of blood cortisol leads to release of Corticotrope Releasing Hormone (CRH) activating the anterior pituitary and signalling the release of Adrenocorticotropic Hormone (ACTH), stimulating the adrenal gland to make more cortisol. In addition to cortisol, the adrenal gland also releases androgen, leading to hyperandrogenism, which gives rise to the symptoms commonly associated with Cortisone Reductase Deficiency.

Genetics

Inactivating mutations in the H6PD gene lead to a lowered supply of NADH, causing cortisone reductase to catalyze the reaction from cortisol to cortisone. This is the most common manifestation of CRD. It has been shown that CRD can be caused by mutations in the HSD11B1 gene as well, specifically mutations caused by K187N and R137C, affecting active site residue and disruption of salt bridges at the subunit interface of the dimer, respectively. In the K187N mutant, activity is abolished, and in the R137C mutant activity is greatly reduced, but not completely abolished.

Symptoms

Cortisol inhibition, and as a result, excess androgen release can lead to a variety of symptoms. Other symptoms come about as a result of increased levels of circulating androgen. Androgen is a steroid hormone, generally associated with development of male sex organs and secondary male sex characteristics The symptoms associated with Cortisone Reductase Deficiency are often linked with Polycystic Ovary Syndrome (PCOS) in females. The symptoms of PCOS include excessive hair growth, oligomenorrhea, amenorrhea, and infertility. In men, cortisone reductase deficiency results in premature pseudopuberty, or sexual development before the age of nine.

Diagnosis and Treatment

Diagnosis of cortisone reductase deficiency is done through analysis of cortisol to cortisone metabolite levels in blood samples. As of now, there is no treatment for cortisone reductase deficiency. Shots of cortisol are quickly metabolised into cortisone by the dysregulated 11β-HSD1 enzyme; however, symptoms can be treated. Treatment of hyperandroginism can be done through prescription of antiandrogens. They do so by inhibiting the release of gonadotropin and luteinizing hormone, both hormones in the pituitary, responsible for the production of testosterone.

Hypophosphatasia

Hypophosphatasia is a rare, and sometimes fatal, metabolic bone disease. Clinical symptoms are heterogeneous, ranging from the rapidly fatal, perinatal variant, with profound skeletal hypomineralization and respiratory compromise, to a milder, progressive osteomalacia later in life. Tissue non-specific alkaline phosphatase (TNSALP) deficiency in osteoblasts and chondrocytes impairs bone mineralization, leading to rickets or osteomalacia. The pathognomonic finding is subnormal serum activity of the TNSALP enzyme, which is caused by one of 200 genetic mutations identified to date, in the gene encoding TNSALP. Genetic inheritance is autosomal recessive for the perinatal and infantile forms but either autosomal recessive or autosomal dominant in the milder forms. The prevalence of hypophosphatasia is not known; one study estimated the live birth incidence of severe forms to be 1:100,000.

Clinical Symptoms

There is a remarkable variety of symptoms that depends, largely, on the age of the patient at initial presentation, ranging from death *in utero* to relatively mild problems with dentition in adult life. Although several clinical sub-types of the disease have been characterized, based on the age at which skeletal lesions are discovered, the disease is best understood as a single continuous spectrum of severity.

Perinatal Hypophosphatasia

Perinatal hypophosphatasia is the most lethal form. Profound hypomineralization results in caput membranaceum (a soft calvarium), deformed or shortened limbs during gestation and at birth, and rapid death due to respiratory failure. Stillbirth is not uncommon and long-term survival is rare. Neonates who manage to survive suffer increasing respiratory compromise due to softening of the bones (osteomalacia) and underdeveloped lungs (hypoplastic). Ultimately, this leads to respiratory failure. Epilepsy (seizures) can occur and can prove lethal. Regions of developing, unmineralized bone (osteoid) may expand and encroach on the marrow space, resulting in myelophthisic anemia.

In radiographic examinations, perinatal hypophosphatasia can be distinguished from even the most severe forms of osteogenesis imperfecta and congenital dwarfism. Some stillborn skeletons show almost no mineralization; others have marked undermineralization and severe osteomalacia. Occasionally, there can be a complete absence of ossification in one or more vertebrae. In the skull, individual bones may calcify only at their centers. Another unusual radiographic feature is bony spurs that protrude laterally from the shafts of the ulnae and fibulae. Despite the considerable patient-to-patient variability and the diversity of radiographic findings, the X-ray can be considered diagnostic.

Infantile Hypophosphatasia

Infantile hypophosphatasia presents in the first 6 months of life, with the onset of poor feeding and inadequate weight gain. Clinical manifestations of rickets often appear at this time. Although cranial sutures appear to be wide, this reflects hypomineralization of the skull, and there is often "functional" craniosynostosis. If the patient survives infancy, these sutures can permanently fuse. Defects in the chest, such as flail chest resulting from rib fractures, lead to respiratory compromise and pneumonia. Elevated calcium in the blood (hypercalcemia) and urine (hypercalcenuria) are also common, and may explain the renal problems and recurrent vomiting seen is this disease.

Radiographic features in infants are generally less severe than those seen in perinatal hypophosphatasia. In the long bones, there is an abrupt change from a normal appearance in the shaft (diaphysis) to uncalcified regions near the ends (metaphysis), which suggests the occurrence of an abrupt metabolic change. In addition, serial radiography studies suggest that defects in skeletal mineralization (i.e. rickets) persist and become more generalized. Mortality is estimated to be 50% in the first year of life.

Childhood Hypophosphatasia

Hypophosphatasia in childhood has variable clinical expression. As a result of defects in the development of the dental cementum, the deciduous teeth (baby teeth) are often lost fore the age of 5. Frequently, the incisors are lost first; occasionally all of the teeth are lost prematurely. Dental radiographs can show the enlarged pulp chambers and root canals that are characteristic of rickets.

Patients may experience delayed walking, a characteristic waddling gait, stiffness and pain, and muscle weakness (especially in the thighs) consistent with nonprogressive myopathy. Typically, radiographs show defects in calcification and characteristic bony defects near the ends of major long bones. Growth retardation, frequent fractures, and low bone density (osteopenia) are common. In severely-affected infants and young children, cranial bones can fuse prematurely, despite the appearance of open fontanels on radiographic studies. The illusion of open fontanels results from hypomineralization of large areas of the calvarium. Premature bony fusion of the cranial sutures may elevate intracranial pressure.

Adult Hypophosphatasia

Adult hypophosphatasia can be associated with rickets, premature loss of deciduous teeth, or early loss of adult dentation followed by relatively good health. Osteomalacia results in painful feet due to poor healing of metatarsal stress fractures. Discomfort in the thighs or hips due to femoral pseudofractures can be distinguished from other types of osteomalacia by their location in the lateral cortices of the femora.

Some patients suffer from calcium pyrophosphate dihydrate crystal depositions with occasional attacks of arthritis (pseudogout), which appears to be the result of elevated endogenous inorganic pyrophosphate (PPi) levels. These patients may also suffer articular cartilage degeneration and pyrophosphate arthropathy. Radiographs reveal pseudofractures in the lateral cortices of the proximal femora and stress fractures, and patients may experience osteopenia, chondrocalcinosis, features of pyrophosphate arthropathy, and calcific periarthritis.

Odontohypophosphatasia is present when dental disease is the only clinical abnormality, and radiographic and/or histologic studies reveal no evidence of rickets or osteomalacia. Although hereditary leukocyte abnormalities and other disorders usually account for this condition, odontohypophosphatasia may explain some "early-onset periodontitis" cases.

Causes

Hypophosphatasia is associated with a molecular defect in the gene encoding tissue non-specific alkaline phosphatase (TNSALP). TNSALP is an enzyme that is tethered to the outer surface of osteoblasts and chondrocytes. TNSALP hydrolyzes several substances, including inorganic pyrophosphate (PPi) and pyridoxal 5'-phosphate (PLP), a major form of vitamin B_6.

When TSNALP is low, inorganic pyrophosphate (PPi) accumulates outside of cells, and inhibits formation of hydroxyapatite, one of the main components of bone, causing rickets in infants and children and osteomalacia (soft bones) in adults. PLP is the principal form of vitamin B_6 and must be dephosphorylated by TNSALP before it can cross the cell membrane. Vitamin B_6 deficiency in the brain impairs synthesis of neurotransmitters, which can cause seizures. In some cases, a build-up of calcium pyrophosphate dihydrate (CPPD) crystals in the joint can cause pseudogout.

Diagnosis

Dental Findings

Hypophosphatasia is often discovered because of an early loss of deciduous (baby or primary) teeth with the root intact. Researchers have recently documented a positive correlation between dental abnormalities and clinical phenotype. Poor dentition is also noted in adults.

Laboratory Testing

The symptom that best characterizes hypophosphatasia is low serum activity of alkaline phosphatase enzyme (ALP). In general, lower levels of enzyme activity correlate with more severe symptoms. The decrease in ALP activity leads to an increase in pyridoxal 5'-phosphate (PLP) in the blood, and correlates with disease severity. Urinary inorganic pyrophosphate (PPi) levels are elevated in most hypophosphatasia patients and, although it remains only a research technique, this increase has been reported to accurately detect carriers of the disease. In addition, most patients have an increased level of urinary phosphoethanolamine (PEA). Tests for serum ALP levels are part of the standard comprehensive metabolic panel (CMP) that is used in routine exams.

Radiography

Despite patient-to-patient variability and the diversity of radiographic findings, the X-ray is diagnostic in infantile hypophosphatasia. Skeletal defects are found in nearly all patients and include hypomineralization, rachitic changes, incomplete vertebrate ossification and, occasionally, lateral bony spurs on the ulnae and fibulae.

In newborns, X-rays readily distinguish hypophosphatasia from osteogenesis imperfecta and congenital dwarfism. Some stillborn skeletons show almost no mineralization; others have marked undermineralization and severe rachitic changes. Occasionally there can be peculiar complete or partial absence of ossification in one or more vertebrae. In the skull, individual membranous bones may calcify only at their centers, making it appear that areas of the unossified calvarium have cranial sutures that are widely separated when, in fact, they are functionally closed. Small protrusions (or "tongues") of radiolucency often extend from the metaphyses into the bone shaft.

In infants, radiographic features of hypophosphatasia are striking, though generally less severe than those found in perinatal hypophosphatasia. In some newly diagnosed patients, there is an abrupt transition from relatively normal-appearing diaphyses to uncalcified metaphases, suggesting an abrupt metabolic change has occurred. Serial radiography studies can reveal the persistence of impaired skeletal mineralization (i.e. rickets), instances of sclerosis, and gradual generalized demineralization.

In adults, X-rays may reveal bilateral femoral pseudofractures in the lateral subtrochanteric diaphysis. These pseudofractures may remain for years, but they may not heal until they break completely or the patient receives intramedullary fixation. These patients may also experience recurrent metatarsal fractures.

Genetic Analysis

All clinical sub-types of hypophosphatasia have been traced to genetic mutations in the gene encoding TNSALP, which is localized on chromosome 1p36.1-34 in humans (ALPL; OMIM#171760). Approximately 204 distinct mutations have been described in the TNSALP gene. An up-to-date list of mutations is available online at The Tissue Nonspecific Alkaline Phosphatase Gene Mutations Database. About 80% of the mutations are missense mutations. The number and diversity of mutations results in highly variable phenotypic expression, and there appears to be a correlation between genotype and phenotype in hypophosphatasia". Mutation analysis is possible and available in 3 laboratories.

Inheritance

Perinatal and infantile hypophosphatasia are inherited as autosomal recessive traits with homozygosity or compound heterozygosity for two defective TNSALP alleles. The mode of inheritance for childhood, adult, and odonto forms of hypophosphatasia can be either autosomal dominant or recessive. Autosomal transmission accounts for the fact that the disease affects males and females with equal frequency. Genetic counseling is complicated by the disease's variable inheritance pattern, and by incomplete penetration of the trait.

Hypophosphatasia is a rare disease that has been reported worldwide and appears to affect individuals of all ethnicities. The prevalence of severe hypophosphatasia is estimated to be 1:100,000 in a population of largely Anglo-Saxon origin. The frequency of mild hypophosphatasia is more challenging to assess because the symptoms may escape notice or be misdiagnosed. The highest incidence of hypophosphatasia has been reported in the Mennonite population in Manitoba, Canada where one in every 25 individuals are considered carriers and one in every 2,500 newborns exhibits severe disease. Hypophosphatasia is considered particularly rare in people of African ancestry in the U.S.

Treatment

As of October 2015, asfotase alfa (Strensiq) has been approved by the FDA for the treatment of hypophosphatasia. Current management consists of palliating symptoms, maintaining calcium balance and applying physical, occupational, dental and orthopedic interventions, as necessary.

- Hypercalcemia in infants may require restriction of dietary calcium or administration of calciuretics. This should be done carefully so as not to increase the skeletal demineralization that results from the disease itself. Vitamin D sterols and mineral supplements, traditionally used for rickets or osteomalacia, should not be used unless there is a deficiency, as blood levels of calcium ions (Ca2+), inorganic phosphate (Pi) and vitamin D metabolites usually are not reduced.

- Craniosynostosis, the premature closure of skull sutures, may cause intracranial hypertension and may require neurosurgical intervention to avoid brain damage in infants.

- Bony deformities and fractures are complicated by the lack of mineralization and impaired skeletal growth in these patients. Fractures and corrective osteotomies (bone cutting) can heal, but healing may be delayed and require prolonged casting or stabilization with orthopedic hardware. A load-sharing intramedullary nail or rod is the best surgical treatment for complete fractures, symptomatic pseudofractures, and progressive asymptomatic pseudofractures in adult hypophosphatasia patients.

- Dental problems: Children particularly benefit from skilled dental care, as early tooth loss can cause malnutrition and inhibit speech development. Dentures may ultimately be needed. Dentists should carefully monitor patients' dental hygiene and use prophylactic programs to avoid deteriorating health and periodontal disease.

- Physical Impairments and pain: Rickets and bone weakness associated with hypophosphatasia can restrict or eliminate ambulation, impair functional endurance, and diminish ability to perform activities of daily living. Nonsteroidal anti-inflammatory drugs may improve pain-associated physical impairment and can help improve walking distance]

- Bisphosphonate (a pyrophosphate synthetic analog) in one infant had no discernible effect on the skeleton, and the infant's disease progressed until death at 14 months of age.

- Bone marrow cell transplantation in two severely affected infants produced radiographic and clinical improvement, although the mechanism of efficacy is not fully understood and significant morbidity persisted.

- Enzyme replacement therapy with normal, or ALP-rich serum from patients with Paget's bone disease, was not beneficial.

- Phase 2 clinical trials of bone targeted enzyme-replacement therapy for the treatment of hypophosphatasia in infants and juveniles have been completed, and a phase 2 study in adults is ongoing.

Discovery

It was discovered initially in 1936 but was fully named and documented by a Canadian Pediatrician, John Campbell Rathbun (1915-1972) while examining and treating a baby boy with very low levels of alkaline phosphatase in 1948. The genetic basis of the disease was mapped out only some 40 years later. This condition is sometimes called "Rathbun's Syndrome" after its principal documenter.

Myeloperoxidase Deficiency

Myeloperoxidase deficiency is an autosomal recessive genetic disorder featuring deficiency, either in quantity or of function, of myeloperoxidase, an enzyme found in certain phagocytic immune cells, especially polymorphonuclear leukocytes.

It can appear similar to chronic granulomatous disease on some screening tests.

Presentation

Although MPO deficiency classically presents with immune deficiency (especially candida albicans infections), the majority of individuals with MPO deficiency show no signs of immunodeficiency.

The lack of severe symptoms suggest that role of myeloperoxidase in the immune response must be redundant to other mechanisms of intracellular killing of phagocytosed bacteria.

Patients with MPO deficiency have a respiratory burst with a normal nitro blue tetrazolium (NBT) test because they still have NADPH oxidase activity, but do not form bleach due to their lack of myeloperoxidase activity. This is in contrast to chronic granulomatous disease, in which the NBT test is 'negative' due to the lack of NADPH oxidase activity (positive test result means neutrophils turn blue, negative means nitroblue tetrazolium remains yellow).

Patients with MPO deficiency are at increased risk for systemic candidiasis.

Neutrophil-Specific Granule Deficiency

Neutrophil-specific granule deficiency (SGD, previously known as lactoferrin deficiency) is a rare congenital immunodeficiency characterized by an increased risk for pyogenic infections due to defective production of specific granules and gelatinase granules in patient neutrophils.

Symptoms

Atypical infections are the key clinical manifestation of SGD. Within the first few years of life, patients will experience repeated pyogenic infections by species such as *Staphylococcus aureus*, *Pseudomonas aeruginosa* or other Enterobacteriaceae, and *Candida albicans*. Cutaneous ulcers or abscesses and pneumonia and chronic lung disease are common. Patients may also develop sepsis, mastoiditis, otitis media, and lymphadenopathy. Infants may present with vomiting, diarrhea, and failure to thrive.

Diagnosis can be made based upon CEBPE gene mutation or a pathognomonic finding of a blood smear showing lack of specific granules. Neutrophils and eosinophils will contain hyposegmented nuclei (a pseudo-Pelger–Huet anomaly).

Genetics

A majority of patients with SGD have been found to have mutations in the CEBPE (CCAAT/enhancer-binding protein epsilon) gene, a transcription factor primarily active in myeloid cells. Almost all patients have been found to be homozygous for the mutation, suggesting the disease is autosomal recessive. One patient, heterozygous for the mutation, was found to be deficient in GFI1, a related gene.

Pathophysiology

The defect in CEBPE appears to block the ability of neutrophils to mature past the promyelocyte stage in bone marrow. Since specific (secondary) and gelatinase (tertiary) granules are only produced past the promyelocyte stage of development, these are deficient in SGD. Lactoferrin is the major enzyme found in specific granules, and will be largely absent in the granulocytes of these patients, along with defensins (despite these also being found in azurophilic (primary) granules). The other major components of azurophilic granules, such as lysozyme, cathepsin, and elastase will be normal, however a lack of defensins and lactoferrin drastically weakens the neutrophil innate ability to fight infection. Neutrophils will also display abnormal chemotaxis, such as a decreased response to fMLP, due to a lack of chemotactic receptors typically found in the specific granules.

Treatment

Treatment consists mainly of high dose antibiotics for active infections and prophylactic antibiotics for prevention of future infections. GM-CSF therapy or bone marrow transplant might be considered for severe cases. Prognosis is difficult to predict, but patients receiving treatment are generally able to survive to adulthood.

Epidemiology

Estimation of the frequency of SGD is difficult, as it is an extremely rare disease with few cases reported in literature. The condition was first reported in 1980, and since only a handful more cases have been published.

Phenylketonuria

Phenylketonuria (PKU) is an inborn error of metabolism involving impaired metabolism of the amino acid phenylalanine. Phenylketonuria is caused by absent or virtually absent phenylalanine hydroxylase (PAH) enzyme activity. Both PKU and most non-PKU hyperphenylalaninemia are the result of phenylalanine hydroxylase deficiency (PAHD).

Protein-rich foods or the sweetener aspartame can act as poisons for people with phenylketonuria. The role of PAH is to break down excess phenylalanine from food. Phenylalanine is a necessary part of the human diet and is naturally present in all kinds of dietary protein. It is also used to make aspartame, known by the trade name Nutrasweet, which is used to sweeten low-calorie and sugar free soft drinks, yogurts, and desserts. In people without PKU, the PAH enzyme breaks down any excess phenylalanine from these sources beyond what is needed by the body. However, if there is not enough of the PAH enzyme or its cofactor, then phenylalanine can build up in the blood and brain to toxic levels, affecting brain development and function. PKU is rare, but important to identify, because if caught early it is very treatable. It is not contagious, and it is lifelong, but with early diagnosis and consistent treatment, the damaging effects can be minimal or non-existent.

Untreated PKU can lead to intellectual disability, seizures, and other serious medical problems. The best proven treatment for classical PKU patients is a strict phenylalanine-restricted diet supplemented by a medical formula containing amino acids and other nutrients. In the United States, the current recommendation is that the PKU diet should be maintained for life. Patients who are diagnosed early and maintain a strict diet can have a normal life span with normal mental development.

PKU is an inherited disease. When an infant is diagnosed with PKU, it is never the result of any action of the parents or any environmental factor. Rather, for a child to inherit PKU, both of his or her parents must have at least one mutated allele of the PAH gene. Most parents who are carriers of PKU genes are not aware that they have this mutation because being a carrier causes no medical problems. To be affected by PKU, a child must inherit two mutated alleles, one from each parent.

Signs and Symptoms

PKU is commonly included in the newborn screening panel of most countries, with varied detection techniques. Most babies in developed countries are screened for PKU soon after birth. Screening for PKU is done with bacterial inhibition assay (Guthrie test), immunoassays using fluorometric or photometric detection, or amino acid measurement using tandem mass spectrometry (MS/MS). Measurements done using MS/MS determine the concentration of Phe and the ratio of Phe to tyrosine, the ratio will be elevated in PKU.

Blood is taken from a two-week-old infant to test for phenylketonuria

Because the mother's body is able to break down phenylalanine during pregnancy, infants with PKU are normal at birth. The disease is not detectable by physical examination at that time, because no damage has yet been done. However, a blood test can reveal elevated phenylalanine levels after one or two days of normal infant feeding. This is the purpose of newborn screening, to detect the disease with a blood test before any damage is done, so that treatment can prevent the damage from happening.

If a child is not diagnosed during the routine newborn screening test (typically performed 2–7 days after birth, using samples drawn by neonatal heel prick), and a phenylalanine restricted diet is not introduced, then phenylalanine levels in the blood will increase over time. Toxic levels of phenylalanine (and insufficient levels of tyrosine) can interfere with infant development in ways which have permanent effects. The disease may present clinically with seizures, hypopigmentation (excessively fair hair and skin), and a "musty odor" to the baby's sweat and urine (due to phenylacetate, a carboxylic acid produced by the oxidation of phenylketone). In most cases, a repeat test should be done at approximately two weeks of age to verify the initial test and uncover any phenylketonuria that was initially missed.

Untreated children often fail to attain early developmental milestones, develop microcephaly, and demonstrate progressive impairment of cerebral function. Hyperactivity, EEG abnormalities, and seizures, and severe learning disabilities are major clinical problems later in life. A characteristic "musty or mousy" odor on the skin, as well as a predisposition for eczema, persist throughout life in the absence of treatment.

The damage done to the brain if PKU is untreated during the first months of life is not reversible. It is critical to control the diet of infants with PKU very carefully so that the brain has an opportunity to develop normally. Affected children who are detected at birth and treated are much less likely to develop neurological problems or have seizures and intellectual disability (though such clinical disorders are still possible.)

In general, however, outcomes for people treated for PKU are good. Treated people may have no detectable physical, neurological, or developmental problems at all. Many adults with PKU who were diagnosed through newborn screening and have been treated since birth have high educational achievement, successful careers, and fulfilling family lives.

Genetics

PKU is an autosomal recessive metabolic genetic disorder. As an autosomal recessive disorder, two PKU alleles are required for an individual to exhibit symptoms of the disease. If both parents are carriers for PKU, there is a 25% chance any child they have will be born with the disorder, a 50% chance the child will be a carrier, and a 25% chance the child will neither develop nor be a carrier for the disease.

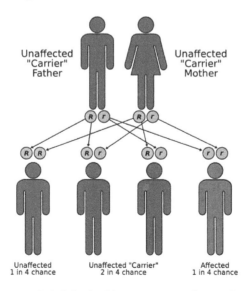

Phenylketonuria is inherited in an autosomal recessive fashion

PKU is characterized by homozygous or compound heterozygous mutations in the gene for the hepatic enzyme phenylalanine hydroxylase (PAH), rendering it nonfunctional. This enzyme is necessary to metabolize the amino acid phenylalanine (Phe) to the amino acid tyrosine (Tyr). When PAH activity is reduced, phenylalanine accumulates and is converted into phenylpyruvate (also known as phenylketone), which can be detected in the urine.

Carriers of a single PKU allele do not exhibit symptoms of the disease but appear to be protected to some extent against the fungal toxin ochratoxin A. This accounts for the persistence of the allele in certain populations in that it confers a selective advantage— in other words, being a heterozygote is advantageous.

The PAH gene is located on chromosome 12 in the bands 12q22-q24.1. More than 400 disease-causing mutations have been found in the PAH gene. This is an example of allelic genetic heterogeneity.

Phenylketonuria can exist in mice, which have been extensively used in experiments into finding an effective treatment for it. The macaque monkey's genome was recently sequenced, and the gene encoding phenylalanine hydroxylase was found to have a sequence that, in humans, would be considered a PKU mutation.

Pathophysiology

Classical PKU

Classical PKU, and its less severe forms "mild PKU" and "mild hyperphenylalaninemia" are caused by a mutated gene for the enzyme phenylalanine hydroxylase (PAH), which converts the amino acid phenylalanine ("Phe") to other essential compounds in the body, in particular tyrosine. Tyrosine is a conditionally essential Amino acid for PKU patients because without PAH it cannot be produced in the body through the breakdown of phenylalanine. Tyrosine is necessary for the production of neurotransmitters like epinephrine, norepinephrine, and dopamine.

PAH deficiency causes a spectrum of disorders, including classic phenylketonuria (PKU) and mild hyperphenylalaninemia (also known as "hyperphe" or "mild HPA"), a less severe accumulation of phenylalanine. Patients with "hyperphe" may have more functional PAH enzyme and be able to tolerate larger amounts of phenylalanine in their diets than those with classic PKU, but unless dietary intake is at least somewhat restricted, their blood Phe levels are still higher than the levels in people with normal PAH activity.

Phenylalanine is a large, neutral amino acid (LNAA). LNAAs compete for transport across the blood–brain barrier (BBB) via the large neutral amino acid transporter (LNAAT). If phenylalanine is in excess in the blood, it will saturate the transporter. Excessive levels of phenylalanine tend to decrease the levels of other LNAAs in the brain.

As these amino acids are necessary for protein and neurotransmitter synthesis, Phe buildup hinders the development of the brain, causing intellectual disability.

Recent research suggests that neurocognitive, psychosocial, quality of life, growth, nutrition, bone pathology are slightly suboptimal even for patients who are treated and maintain their Phe levels in the target range, if their diet is not supplemented with other amino acids.

Classic PKU dramatically affects myelination and white matter tracts in untreated infants; this may be one major cause of neurological disorders associated with phenylketonuria. Differences in white matter development are observable with magnetic resonance imaging. Abnormalities in gray matter can also be detected, particularly in the motor and pre-motor cortex, thalamus and the hippocampus.

It was recently suggested that PKU may resemble amyloid diseases, such as Alzheimer's disease and Parkinson's disease, due to the formation of toxic amyloid-like assemblies of phenylalanine.

Other non-PAH mutations can also cause PKU.

Tetrahydrobiopterin-Deficient Hyperphenylalaninemia

A rarer form of hyperphenylalaninemia occurs when the PAH enzyme is normal, and a defect is found in the biosynthesis or recycling of the cofactor tetrahydrobiopterin (BH_4). BH_4 (called biopterin) is necessary for proper activity of the enzyme PAH, and this coenzyme can be supplemented as treatment. Those who suffer from this form of hyperphenylalaninemia may have a deficiency of tyrosine (which is created from phenylalanine by PAH). These patients must be supplemented with tyrosine to account for this deficiency.

Dihydrobiopterin reductase activity is needed to replenish quinonoid-dihydrobiopterin back into its tetrahydrobiopterin form, which is an important cofactor in many reactions in amino acid metabolism. Those with this deficiency may produce sufficient levels of the enzyme phenylalanine hydroxylase (PAH) but, since tetrahydrobiopterin is a cofactor for PAH activity, deficient dihydrobiopterin reductase renders any PAH produced unable to use phenylalanine to produce tyrosine. Tetrahydrobiopterin is a cofactor in the production of L-DOPA from tyrosine and 5-hydroxy-L-tryptophan from tryptophan, which must be supplemented as treatment in addition to the supplements for classical PKU.

Levels of dopamine can be used to distinguish between these two types. Tetrahydrobiopterin is required to convert Phe to Tyr and is required to convert Tyr to L-DOPA via the enzyme tyrosine hydroxylase. L-DOPA, in turn, is converted to dopamine. Low levels of dopamine lead to high levels of prolactin. By contrast, in classical PKU (without dihydrobiopterin involvement), prolactin levels would be relatively normal.

Tetrahydrobiopterin deficiency can be caused by defects in four genes. They are known as HPABH4A, HPABH4B, HPABH4C, and HPABH4D.

Metabolic Pathways

The enzyme phenylalanine hydroxylase normally converts the amino acid phenylalanine into the amino acid tyrosine. If this reaction does not take place, phenylalanine accumulates and tyrosine is deficient. Excessive phenylalanine can be metabolized into phenylketones through the minor route, a transaminase pathway with glutamate. Metabolites include phenylacetate, phenylpyruvate and phenethylamine. Elevated levels of phenylalanine in the blood and detection of phenylketones in the urine is diagnostic, however most patients are diagnosed via newborn screening.

Pathophysiology of phenylketonuria, which is due to the absence of functional phenylalanine hydroxylase (classical subtype) or functional enzymes for the recycling of tetrahydrobiopterin (new variant subtype) utilized in the first step of the metabolic pathway.

Treatment

PKU is not curable. However, if PKU is diagnosed early enough, an affected newborn can grow up with normal brain development by managing and controlling phenylalanine ("Phe") levels through diet, or a combination of diet and medication.

When Phe cannot be metabolized by the body, a typical diet that would be healthy for people without PKU causes abnormally high levels of Phe to accumulate in the blood, which is toxic to the brain. If left untreated, complications of PKU include severe intellectual disability, brain function abnormalities, microcephaly, mood disorders, irregular motor functioning, and behavioral problems such as attention deficit hyperactivity disorder, as well as physical symptoms such as a "musty" odor, eczema, and unusually light skin and hair coloration. In contrast, PKU patients who follow the prescribed dietary treatment from birth, may have no symptoms at all. Their PKU would be detectable only by a blood test.

To achieve these good outcomes, all PKU patients must adhere to a special diet low in Phe for optimal brain development. Since Phe is necessary for the synthesis of many proteins, it is required for appropriate growth, but levels must be strictly controlled in PKU patients.

PKU is not a food allergy or a digestive problem. Eating "forbidden" foods does not cause an immediate reaction. However, some people with PKU can be very sensitive to quick changes in Phe levels causing a "Protein High". Quick changes can be caused by absorption of PKU formula causing the level of phenylalanine to drop suddenly or a quick rise in Phe level after a meal high in protein. The phenylalanine from that food remains in the person's system, however, and as Phe accumulates over time they may experience concentration, confusion and mood problems, as well as eczema and other symptoms. For children, developmental problems may occur if levels are elevated frequently or remain elevated for a significant amount of time. Changes in Phe levels, while sleeping at night, can prevent some people with PKU from getting enough rest, making it more difficult to concentrate during the day.

Optimal health ranges (or "target ranges") are between 120 and 360 µmol/L or equivalently 2 to 6 mg/dL, and aimed to be achieved during at least the first 10 years, to allow the brain to develop normally.

In the past, PKU-affected people were allowed to go off diet after approximately eight, then 18 years of age. Today, most physicians recommend PKU patients must manage their Phe levels throughout life. For teens and adults, somewhat higher levels of Phe may be tolerable, but restriction is still advised to prevent mood disorders and difficulty concentrating, among other neurological problems.

The diet requires severely restricting or eliminating foods high in Phe, such as soybeans, seal meat, egg whites, shrimps, chicken breast, spirulina, watercress, fish, whale, nuts, crayfish, lobster, tuna, turkey, legumes, elk meat and lowfat cottage cheese. Starchy foods, such as potatoes and corn are generally acceptable in controlled amounts, but the quantity of Phe consumed from these foods must be monitored. A food diary is usually kept to record the amount of Phe consumed with each meal, snack, or drink. An "exchange" system can be used to calculate the amount of Phe in a food from the protein content identified on a nutritional information label. Lower-protein "medical food" substitutes are often used in place of normal bread, pasta, and other grain-based foods, which contain a significant amount of Phe. Many fruits and vegetables are lower in Phe and can be eaten in larger quantities. Infants may still be breastfed to provide all of the benefits of breastmilk, but the quantity must also be monitored and supplementation for missing nutrients will be required. The sweetener aspartame, present in many diet foods and soft drinks, must also be avoided, as aspartame contains phenylalanine.

Different patients can tolerate different amounts of Phe in their diet. Regular blood tests are required to determine the effects of dietary Phe intake on blood Phe level.

Patients typically work with a professional dietitian to find a diet that meets their nutritional needs without causing their blood Phe level to exceed the target range.

Supplementary "protein substitute" formulas are typically prescribed for Classical PKU patients (starting in infancy) to provide the amino acids and other necessary nutrients that would otherwise be lacking in a low-phenylalanine diet. Tyrosine, which is normally derived from phenylalanine and which is necessary for normal brain function, is usually supplemented. Consumption of the protein substitute formulas can actually reduce phenylalanine levels, probably because it stops the process of protein catabolism from releasing Phe stored in the muscles and other tissues into the blood. Many PKU patients have their highest Phe levels after a period of fasting (such as overnight), because fasting triggers catabolism. A diet that is low in phenylalanine but does not include protein substitutes may also fail to lower blood Phe levels, since a nutritionally insufficient diet may also trigger catabolism. For all these reasons, the prescription formula is an important part of the treatment for patients with classic PKU.

The oral administration of tetrahydrobiopterin (or BH4) (a cofactor for the oxidation of phenylalanine) can reduce blood levels of this amino acid in certain patients. The company BioMarin Pharmaceutical has produced a tablet preparation of the compound sapropterin dihydrochloride (Kuvan), which is a form of tetrahydrobiopterin. Kuvan is the first drug that can help BH4-responsive PKU patients (defined among clinicians as about 1/2 of the PKU population) lower Phe levels to recommended ranges. Working closely with a dietitian, some PKU patients who respond to Kuvan may also be able to increase the amount of natural protein they can eat. After extensive clinical trials, Kuvan has been approved by the FDA for use in PKU therapy. Some researchers and clinicians working with PKU are finding Kuvan a safe and effective addition to dietary treatment and beneficial to patients with PKU.

Biomarin is currently conducting clinical trials to investigate another type of treatment for PKU. PEG-PAL (PEGylated recombinant phenylalanine ammonia lyase or 'PAL') is an enzyme substitution therapy in which the missing PAH enzyme is replaced with an analogous enzyme that also breaks down Phe. PEG-PAL is now in Phase 2 clinical development to treat patients who do not adequately respond to KUVAN.

Dietary supplementation with large neutral amino acids(LNAAs), with or without the traditional PKU diet is another treatment strategy. The LNAAs (e.g. leu, tyr, trp, met, his, ile, val, thr) compete with phe for specific carrier proteins that transport LNAAs across the intestinal mucosa into the blood and across the blood brain barrier into the brain .

Studies have demonstrated that PKU patients given daily supplements of LNAAs have decreased plasma phe levels and reduced brain phe concentrations measured by magnetic resonance spectroscopy.

Another interesting treatment strategy for PKU patients is casein glycomacropeptide (CGMP), which is a milk peptide naturally free of Phe in its pure form CGMP can substitute the main part of the free amino acids in the PKU diet and provides several beneficial nutritional effects compared to free amino acids. The fact that CGMP is a peptide ensures that that the absorption rate of its amino acids is prolonged compared to free amino acids and thereby results in improved protein retention and increased satiety compared to free amino acids. Another important benefit of CGMP is that the taste is significantly improved when CGMP substitutes part of the free amino acids and this may help ensure improved compliance to the PKU diet.

Furthermore, CGMP contains a high amount of the phe lowering LNAAs, which constitutes about 41 g per 100 g protein and will therefore help maintain plasma phe levels in the target range.

Other therapies are currently under investigation, including gene therapy.

Maternal Phenylketonuria

For women with phenylketonuria, it is essential for the health of their children to maintain low Phe levels before and during pregnancy. Though the developing fetus may only be a carrier of the PKU gene, the intrauterine environment can have very high levels of phenylalanine, which can cross the placenta. The child may develop congenital heart disease, growth retardation, microcephaly and intellectual disability as a result. PKU-affected women themselves are not at risk of additional complications during pregnancy.

In most countries, women with PKU who wish to have children are advised to lower their blood Phe levels (typically to between 2 and 6 mg/dL) before they become pregnant, and carefully control their levels throughout the pregnancy. This is achieved by performing regular blood tests and adhering very strictly to a diet, in general monitored on a day-to-day basis by a specialist metabolic dietitian. In many cases, as the fetus' liver begins to develop and produce PAH normally, the mother's blood Phe levels will drop, requiring an increased intake to remain within the safe range of 2–6 mg/dL. The mother's daily Phe intake may double or even triple by the end of the pregnancy, as a result. When maternal blood Phe levels fall below 2 mg/dL, anecdotal reports indicate that the mothers may suffer adverse effects, including headaches, nausea, hair loss, and general malaise. When low phenylalanine levels are maintained for the duration of pregnancy, there are no elevated levels of risk of birth defects compared with a baby born to a non-PKU mother.

Epidemiology

The average number of new cases of PKU varies in different human populations. United States Caucasians are affected at a rate of 1 in 10,000. Turkey has the highest documented rate in the world, with 1 in 2,600 births, while countries such as Finland and Japan have extremely low rates with fewer than one case of PKU in 100,000 births. A

1987 study from Slovakia reports a Roma population with an extremely high incidence of PKU (one case in 40 births) due to extensive inbreeding. It is the most common amino acid metabolic problem in the United Kingdom.

Country	Incidence of PKU
Australia	1 in 10,000
Canada	1 in 22,000
China	1 in 17,000
Czechoslovakia	1 in 7,000
Denmark	1 in 12,000
Finland	1 in 200,000
France	1 in 13,500
India	1 in 18,300
Ireland	1 in 4,500
Italy	1 in 17,000
Japan	1 in 125,000
Korea	1 in 41,000
Norway	1 in 14,500
Turkey	1 in 2,600
Philippines	1 in 102,000
Scotland	1 in 5,300
United Kingdom	1 in 14,300
United States	1 in 15,000

History

Before the causes of PKU were understood, PKU caused severe disability in most people who inherited the relevant mutations. Nobel and Pulitzer Prize winning author Pearl S. Buck had a daughter named Carol who lived with PKU before treatment was available, and wrote a moving account of its effects in a book called *The Child Who Never Grew*. Many untreated PKU patients born before widespread newborn screening are still alive, largely in dependent living homes/institutions.

Phenylketonuria was discovered by the Norwegian physician Ivar Asbjørn Følling in 1934 when he noticed hyperphenylalaninemia (HPA) was associated with intellectual disability. In Norway, this disorder is known as Følling's disease, named after its discoverer. Dr. Følling was one of the first physicians to apply detailed chemical analysis to the study of disease.

In 1934 at Rikshospitalet, Dr. Følling saw a young woman named Borgny Egeland. She had two children, Liv and Dag, who had been normal at birth but subsequently developed intellectual disability. When Dag was about a year old, the mother noticed a strong smell to his urine. Dr. Følling obtained urine samples from the children and, after many tests, he found that the substance causing the odor in the urine was phenylpyruvic acid. The children, he concluded, had excess phenylpyruvic acid in the urine, the condition which came to be called phenylketonuria (PKU).

His careful analysis of the urine of the two affected siblings led him to request many physicians near Oslo to test the urine of other affected patients. This led to the discovery of the same substance he had found in eight other patients. He conducted tests and found reactions that gave rise to benzaldehyde and benzoic acid, which led him to conclude that the compound contained a benzene ring. Further testing showed the melting point to be the same as phenylpyruvic acid, which indicated that the substance was in the urine. His careful science inspired many to pursue similar meticulous and painstaking research with other disorders.

PKU was the first disorder to be routinely diagnosed through widespread newborn screening. Robert Guthrie introduced the newborn screening test for PKU in the early 1960s. With the knowledge that PKU could be detected before symptoms were evident, and treatment initiated, screening was quickly adopted around the world. Austria started screening for PKU in 1966 and England in 1968.

Pseudocholinesterase Deficiency

Pseudocholinesterase deficiency is an inherited blood plasma enzyme abnormality in which the body's production of butyrylcholinesterase (BCHE; pseudocholinesterase) is impaired. People who have this abnormality may be sensitive to certain anesthetic drugs, including the muscle relaxants succinylcholine and mivacurium as well as other ester local anesthetics.

Affected Groups

Arya Vysyas

Multiple studies done both in and outside India have shown an increased prevalence of pseudocholinesterase deficiency amongst the Arya Vysya community. A study performed in the Indian State of Tamil Nadu in Coimbatore, on 22 men and women from this community showed that 9 of them had pseudocholinesterase deficiency, which translates to a prevalence that is 4000-fold higher than that in European and American populations.

Persian Jews

Pseudocholinesterase deficiency (anesthesia sensitivity) is an autosomal recessive condition common within the Persian and Iraqi Jewish populations. Approximately one in 10 Persian Jews are known to have a mutation in the gene causing this disorder and thus one in 100 couples will both carry the mutant gene and each of their children will have a 25% chance of having two mutant genes, and thus be affected with this disorder. This means that one out of 400 Persian Jews is affected with this condition.

Effects

The effects are varied depending on the particular drug given. When anesthetists administer standard doses of these anesthetic drugs to a person with pseudocholinesterase deficiency, the patient experiences prolonged paralysis of the respiratory muscles, requiring an extended period of time during which the patient must be mechanically ventilated. Eventually the muscle-paralyzing effects of these drugs will wear off despite the deficiency of the pseudocholinesterase enzyme. If the patient is maintained on a mechanical respirator until normal breathing function returns, there is little risk of harm to the patient.

However, because it is rare in the general population, it is sometimes overlooked when a patient does not wake-up after surgery. If this happens, there are two major complications that can arise. First, the patient may lie awake and paralyzed, while medical providers try to determine the cause of the patient's unresponsiveness. Second, the breathing tube may be removed before the patient is strong enough to breathe properly, potentially causing respiratory arrest.

This enzyme abnormality is a benign condition unless a person with pseudocholinesterase deficiency is exposed to the offending pharmacological agents.

Complications

The main complication resulting from pseudocholinesterase deficiency is the possibility of respiratory failure secondary to succinylcholine or mivacurium-induced neuromuscular paralysis.

Individuals with pseudocholinesterase deficiency also may be at increased risk of toxic reactions, including sudden cardiac death, associated with recreational use of cocaine.

Prognosis

Prognosis for recovery following administration of succinylcholine is excellent when medical support includes close monitoring and respiratory support measures.

In nonmedical settings in which subjects with pseudocholinesterase deficiency are exposed to cocaine, sudden cardiac death can occur.

Patient Education

Patients with known pseudocholinesterase deficiency may wear a medic-alert bracelet that will notify healthcare workers of increased risk from administration of succinylcholine.

These patients also may notify others in their family who may be at risk for carrying one or more abnormal pseudocholinesterase gene alleles.

Drugs to avoid

Drugs Containing Succinylcholine - e.g. Quelicin & Anectine

These drugs are commonly given as muscle relaxants prior to surgery. That means that victims of this deficiency cannot receive certain anesthetics.

A dose that would paralyze the average individual for 3 to 5 mins can paralyze the enzyme-deficient individual for up to 2 hours. The neuro-muscular paralysis can go on for up to 8 hours.

If this condition is recognized by the anesthesiologist early, then there is rarely a problem. Even if the patient is given succinylcholine, he can be kept intubated and sedated until the muscle relaxation resolves.

Drugs Containing Mivacurium - e.g. Mivacron

Mivacron is also a muscle relaxant that is used prior to inserting a tube for breathing.

Drugs Containing Pilocarpine - e.g. Salagen

Salagen is used to treat dry mouth. As the name suggests, dry mouth is a medical condition that occurs when saliva production goes down. There are lots of different causes of dry mouth including side effect of various drugs.

Drugs Containing Butyrylcholine

Use of butyrylcholine is not common. It can be used to treat exposure to nerve agents, pesticides, toxins, etc.

Drugs Containing Huperzine A and Donepezil

These drugs are used to slow the progression of Alzheimer's disease.

Drugs Containing Propionylcholine and Acetylcholine

Drugs containing Parathion

Parathion is used as an agricultural pesticide. Exposure to pesticides with Parathion should be avoided.

Procaine drugs e.g. Novocaine

This drug is injected before and during various surgical or dental procedures or labor and delivery. Procaine causes loss of feeling in the skin and surrounding tissues.

Testing

This inherited condition can be diagnosed with a blood test. If the total cholinesterase activity in the patient's blood is low, this may suggest an atypical form of the enzyme is present, putting the patient at risk of sensitivity to suxamethonium and related drugs. Inhibition studies may also be performed to give more information about potential risk. In some cases, genetic studies may be carried out to help identify the form of the enzyme that is present.

References

- Rapini, Ronald P.; Bolognia, Jean L.; Jorizzo, Joseph L. (2007). Dermatology: 2-Volume Set. St. Louis: Mosby. ISBN 1-4160-2999-0.

- James, William D.; Berger, Timothy G.; et al. (2006). Andrews' Diseases of the Skin: clinical Dermatology. Saunders Elsevier. ISBN 0-7216-2921-0.

- Kasper, DL, Fauci, AS, Longo, DL, Braunwald, E, Hauser, SL, and Jameson, JL. Harrison's Principles of Internal Medicine, 16th edition 2005;357. ISBN 0-07-139140-1.

- Whyte MP (2001). "Hypophosphatasia". In Scriver CR, Beaudet AL, Sly WS, Valle D, Vogelstein B. The Metabolic & Molecular Bases of Inherited Disease. 4 (8th ed.). New York: McGraw-Hill. pp. 5313–29. ISBN 0-07-913035-6.

- James, William D.; Berger, Timothy G. (2006). Andrews' Diseases of the Skin: clinical Dermatology. Saunders Elsevier. ISBN 0-7216-2921-0.

- Chapter 55, page 255 in:Behrman, Richard E.; Kliegman, Robert; Nelson, Waldo E.; Karen Marcdante; Jenson, Hal B. (2006). Nelson essentials of pediatrics. Elsevier/Saunders. ISBN 1-4160-0159-X.

- Sample, Iian (29 February 2012). "Gene therapy cures life-threatening lung infection in teenage boy". The Guardian. Retrieved 6 November 2015.

- Cedars-Sinai Medical Genetics Institute. (2009). "Genetic Screening in the Persian Jewish Community". [1] Retrieved July 20, 2011.

Permissions

We would like to thank the editorial team for lending their expertise to make the book truly unique. They have played a crucial role in the development of this book. Without their invaluable contributions this book wouldn't have been possible. They have made vital efforts to compile up to date information on the varied aspects of this subject to make this book a valuable addition to the collection of many professionals and students.

This book was conceptualized with the vision of imparting up-to-date and integrated information in this field. To ensure the same, a matchless editorial board was set up. Every individual on the board went through rigorous rounds of assessment to prove their worth. After which they invested a large part of their time researching and compiling the most relevant data for our readers.

The editorial board has been involved in producing this book since its inception. They have spent rigorous hours researching and exploring the diverse topics which have resulted in the successful publishing of this book. They have passed on their knowledge of decades through this book. To expedite this challenging task, the publisher supported the team at every step. A small team of assistant editors was also appointed to further simplify the editing procedure and attain best results for the readers.

Apart from the editorial board, the designing team has also invested a significant amount of their time in understanding the subject and creating the most relevant covers. They scrutinized every image to scout for the most suitable representation of the subject and create an appropriate cover for the book.

The publishing team has been an ardent support to the editorial, designing and production team. Their endless efforts to recruit the best for this project, has resulted in the accomplishment of this book. They are a veteran in the field of academics and their pool of knowledge is as vast as their experience in printing. Their expertise and guidance has proved useful at every step. Their uncompromising quality standards have made this book an exceptional effort. Their encouragement from time to time has been an inspiration for everyone.

The publisher and the editorial board hope that this book will prove to be a valuable piece of knowledge for students, practitioners and scholars across the globe.

Index

www.ingramcontent.com/pod-product-compliance
Lightning Source LLC
Jackson TN
JSHW052200130125
77033JS00004B/197